AF885346

BUSH HISTORIES *I*
Gold-Seekers in Greenstone & Beyond

ALSO BY E.J. LAVOIE

Nonfiction

. . . And the Geraldton Way (1987)
Muskeg Tours (1987)
Logging in Longlac (1989)
Steel Lake Canoe Trail (2002)
The March of Times Star - Book I: 1945-1965 (2009)
Muskeg Tours 2009 (2009) [eBook]
Greenstone's Believe It or Not!:
101 Incredible Facts About Our History - plus images (2015)

BUSH HISTORIES *I*
Gold-Seekers in Greenstone & Beyond

E.J. Lavoie

WHISKYJACK PUBLISHING
Greenstone * Canada

Copyright © 2016 Edgar J. Lavoie

Edgar J. Lavoie writing as E.J. Lavoie.

All rights reserved.

Library and Archives Canada Cataloguing in Publication

Lavoie, Edgar J. (Edgar Joseph), 1940-, author
Bush histories / E.J. Lavoie.
Volume I based on the Muskeg tours 2009, published in DVD format.
Contents: Volume I. Gold-seekers in Greenstone & beyond.
ISBN 978-0-9940196-0-8 (v. 1 : paperback)
1. Mineral industries--Ontario, Northern--History. 2. Mines and mineral resources--Ontario, Northern--History. I. Title.
HD9506.C23O55 2016 338.209713'1 C2016-905812-3

Original Edition

The following book is a reprint of the electronic version of ***Muskeg Tours 2009 : Explorations in the Beardmore-Geraldton Gold Camp***, published on DVD in 2009 under the business name of Experience North. What follows is a facsimile edition with the original formatting. Hereinafter, the reader must re-orient the book to follow the pages in landscape formatting. All content after this page will reference the original edition.

Disclaimer

Advertising material was submitted by the individual businesses or groups. Personal or official remarks have been attributed. In the interests of historical accuracy, advertising and remarks have been reproduced. The author accepts responsibility for all reportage.

Typeset in Times New Roman by WhiskyJack Publishing, Canada.
Printed and bound in the United States of America.

WhiskyJack Publishing, PO Box 279, Geraldton, ON, Canada P0T 1M0
www.WhiskyJackPublishing.ca

Print books available in black-and-white version or full colour version.
E-book version available at: Amazon.com
Follow the author on his blog: http://ejlavoie.wordpress.com

FACSIMILE EDITION

Muskeg Tours 2009

Print Version
FREE

Explorations in the Beardmore-Geraldton Gold Camp

Edgar J. Lavoie

Greenstone, Ontario

Experience North

Muskeg Tours 2009

Explorations in the Beardmore-Geraldton Gold Camp

Edgar J. Lavoie

Chibougamau Drilling

G C F

Société du développement économique de GREENSTONE Economic Development Corporation

KODIAK

MPP

Municipality of Greenstone

NORONT RESOURCES LTD.

NOAC

NSD

PREMIER Gold Mines Limited

ROXMARK MINES LIMITED

SAGE GOLD INC.

SPIDER

Greenstone, Ontario EXPERIENCE NORTH 2009

Print Version
Free Distribution

Front Cover Photo: Stretched pillow lava in the Gold Strike Zone, West Geraldton Property, of Kodiak Exploration Ltd. Note quartz-tourmaline veins and glacial striations.
Back Cover Photo: This historic photo replicates most of the front cover image of the original **Muskeg Tours**, published in 1987. The photo of the Little Long Lac mine was snapped from the Barton Bay Bridge on May 2, 1937

ALSO BY EDGAR J. LAVOIE
And the Geraldton Way . . .
Muskeg Tours (1987)
The March of Times Star, Book 1

This book is a desktop publication of:
Experience North
Edgar J. Lavoie, Proprietor
PO Box 205
Geraldton, ON
P0T 1M0

http://www.goshencanada.ca

Telephone orders 1-807-854-1184
Fax orders 1-807-854-2289
E-mail orders edgarlavoie@hotmail.com

© 2009 - Digital & Print Versions
All rights reserved.

Sources of Maps and Photographs
Maps from OGS/Ministry of Northern Development & Mines
Historic images from various sources.
Photos without credit lines are attributed to the author.

By special arrangement, the publisher can supply a CD with a print version that can be sold. Sale of this print version is a contravention of copyright.

Canadian Cataloging in Publication Data
Lavoie, Edgar J., 1940-
Muskeg Tours 2009: explorations in the Beardmore-Geraldton gold camp. Includes table of contents and introductory material.
ISBN 978-0-9809017-1-9
1. General History of North America – Canada – Ontario I. Title

Digital version produced in Canada by Experience North.

By arrangement, the publisher can supply updated and customized digital versions .

CONTENTS

MESSAGE FROM THE MAYOR

The Beardmore-Geraldton Gold Camp was a world - renowned producer of gold for many decades in the past. The last mine in the Geraldton area of the camp ceased production in 1970. Since then, there have been sporadic attempts at exploration by following the old mining adage, "Look for gold in the shadow of a headframe."

The original ***Muskeg Tours*** publication built on this knowledge to improve tourism potential. Previous books and publications have supplied the history of this gold camp, and mapped out modern-day exploration "tours".

New technology and the high price of gold have provided a new incentive for exploration in this area. A number of exploration companies have expended millions of dollars to identify the location and extent of economic mineral deposits that will result in new mines coming into production in the Beardmore-Geraldton Camp.

This camp is located in the Municipality of Greenstone. Exploration has already provided a significant economic benefit to a region that has been devastated by the downturn in the forestry sector.

Mining was the reason for the flourishing of many communities in this camp, and mining will be the main reason for the resurgence of these communities.

Mining, forestry, and tourism will be the three pillars of the renewal of economic prosperity in Greenstone. As the Mayor of Greenstone, it is my pleasure to welcome you to the future, and to invite you to come and visit with us as the excitement builds.

Mayor Michael Power
Municipality of Greenstone
February, 2009

REMARKS OF THE MINISTER

These are exciting times for the mineral development industry throughout Northwestern Ontario. More than 300 active exploration programs are being conducted by prospectors and by junior and major exploration companies in the region. The region is truly a hotbed of mineral development activity within a province that still leads the nation in exploration projects and expenditures.

The Beardmore-Geraldton gold camp is part of that excitement. In an area with a glorious and productive mining past, recent discoveries have made it a renewed focal point for mineral developers. It is now considered to be one of the most prospective gold camps in the Canadian Shield.

Our government seeks to build on this exceptional activity, along with our rich mineral endowment and our knowledge and expertise in all areas of mining, to create sustainable growth. Guided by Ontario's Mineral Development Strategy, we are building on these considerable strengths to ensure that our industry remains economically, socially, and environmentally sustainable.

Our government remains committed to supporting mineral sector excellence in areas like Beardmore-Geraldton, and to supporting an industry recognized as a global leader and an essential contributor to our provincial prosperity.

The Honourable Michael Gravelle
Ontario Minister of Northern
Development and Mines
February, 2009

PREFACE OF THE ONTARIO GEOLOGICAL SURVEY

The first documented mineral exploration activity in the Greenstone area dates back to gold prospecting in the 1880's. The turn of the 20th century was dominated by iron exploration, but gold became the metal sought during "rushes" in 1915 (Tashota), 1925 (Beardmore), and 1931 (Geraldton). Prospectors made exciting new gold discoveries and mines quickly followed.

The Beardmore-Geraldton Greenstone Belt would eventually produce 4.1 million ounces of gold and over 300,000 ounces of silver between 1934 and 1970. Most of this production came from two gold camps at Beardmore and Geraldton. The balance of gold production came from several small mines scattered northeast of Beardmore and one mine south of Longlac.

Although the last gold mine closed in 1970, the camp experienced renewed interest following unprecedented rises in gold prices, which until that time had remained fixed by governments. Levels of exploration activity over the next three decades rose and fell in concert with gold price. Increasing worldwide demand for metals led to the onset of an exploration "super-cycle" in 2005. Companies were able to secure venture capital from stock markets, leading to the acquisition and exploration of record numbers of mining claims throughout Ontario.

Fueled by the discovery of the Golden Mile gold zone on the Hercules property by Kodiak Exploration Limited in mid-2007, the belt experienced yet another staking rush. As a result, all prospective ground in the belt along a 150 km-long corridor from Lake Nipigon to east of Longlac was staked by Kodiak and a dozen other companies. It is estimated that these companies spent in excess of $35M in the Greenstone area in 2008.

While exciting new discoveries continue to be made, it is important to note that many of their properties are in the shadow of past-producing mines. These properties are being re-evaluated in light of new ideas and concepts that reflect our growing and ever-changing understanding of geology and mineral deposits.

The first geological survey of the area was undertaken by the Geological Survey of Canada in 1869. Since 1906, the Ontario Geological Survey

(OGS) and its predecessors – the Ontario Bureau of Mines and the Ontario Department of Mines – have had a longstanding relationship with the Greenstone area and its prospectors, explorationists, and miners. The OGS continues to provide high-quality, useful information about the area's geology and mineral deposits in the hope that a new chapter will soon be added to Greenstone's storied mining heritage.

We applaud Edgar Lavoie in his efforts to bring that heritage to light.

Mark Smyk, P.Geo
Regional Resident Geologist, Thunder Bay
North District

Gerry White, P.Geo.
District Geologist, Thunder Bay North District

John Mason, P.Geo
Northwest Regional Manager, Resident
Geologist Program
Ontario Geological Survey
Ministry of Northern Development and Mines
February, 2009

FOREWORD OF THE AUTHOR

The stories in this volume are addressed to a general audience. If you are an insider in the mining industry, and expecting to read highly technical reports, you may be disappointed.

However, everybody loves a good story. Especially when pains have been taken to get the technical details right. So prepare yourself for a good read. The stories, by the way, have been so well received in large part because of the cooperation of the organizations and the individuals who were the subjects of the stories.

The stories were published over a two-year period in three publications. Most of them appeared in Geraldton's community weekly, the *Times Star*. A few others appeared in Thunder Bay's regional daily, *The Chronicle-Journal*, and in Sudbury's province-wide monthly publication, *Northern Ontario Business*.

This volume also contains original, previously unpublished text, images, graphics, and maps. For the maps, unless otherwise indicated, credit goes to the Thunder Bay Resident Geologist Program of the Ontario Geological Survey, Ministry of Northern Development and Mines.

Now a few words about the book's format. The book may be read on a computer monitor, or it may be printed out. The different sections of the book are presented in separate PDF's, for good reason. You may wish to print out only selected sections. And, the format makes it possible to re-issue the book in successive editions, with updates.

Pages – even the entire volume – may be collected in a loose-leaf binder. For professional binding, consult a print shop or bookbinder.

All errors in editing are my fault. I apologize.

I wish to thank the people – too many to list, for I'd inevitably forget some – who encouraged the publication of this volume, and I express my gratitude to the sponsors, who provided financial support.

Edgar J. Lavoie
Author/Editor/Publisher
February, 2009

A Technical Note: The book is composed in Microsoft Word. Pages may be printed on letter-size paper. The front and back covers should be printed on legal-size paper and then trimmed to letter-size. Also, a glossy card stock may be more suitable for the covers.

Year 2007 :
Chapters 1 to 7

Muskeg Tours 2009

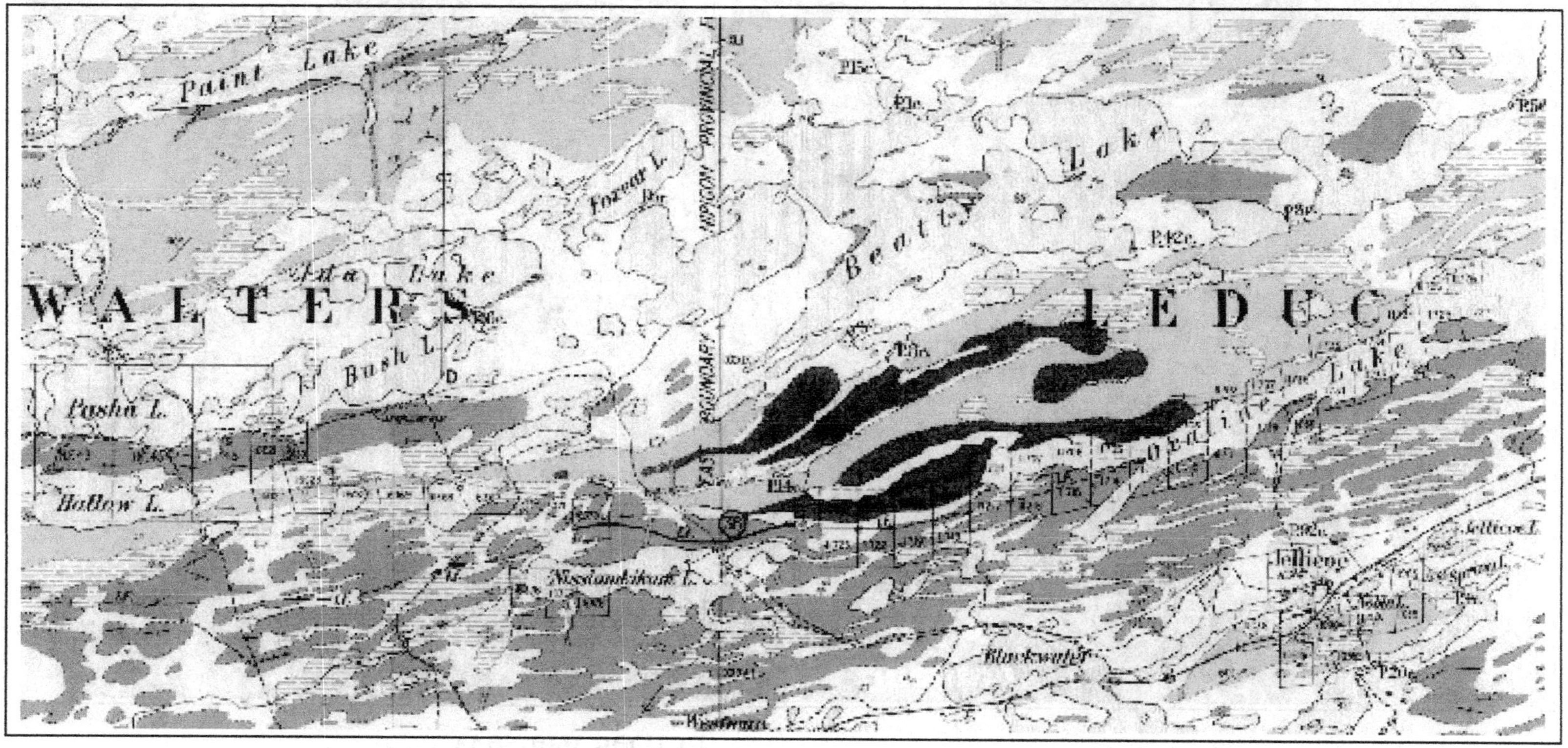

Section of Map No. 45a, "Sturgeon River Gold Area", ODM 1936

BEARDMORE-GERALDTON GOLD BELT : Western Section (Google Earth View From 40 Miles Altitude)

STURGEON RIVER GOLD CAMP, 1936 :

Beardmore-Geraldton Gold Belt

In 1914 the Canadian Northern Ontario Railway finished its line between Sudbury and Port Arthur (now Thunder Bay North). Jellicoe, between Longlac and Nipigon, became a divisional point. Spotted along the line were numerous watering stops, with names such as Nezah and Jackpine and Beardmore. Of course, after leaving the railway, prospectors still resorted to their traditional transport, the canoe.

Not much happened east of Lake Nipigon until the discovery of gold near Beardmore in 1925. The Northern Empire mine developed beside the Blackwater River and railway line (by then renamed the Canadian National Railway). By 1934 Northern Empire Mines Company, Limited had sunk a shaft to a depth of 609 feet, had completed a 125-ton mill, and had poured its first brick. In 1936 the mine produced 28,575 ounces of gold and 3,756 ounces of silver.

There was no road link to the rest of Canada. A hydro line serviced the mine, bringing power along the CNR right-of-way from Cameron Falls on the Nipigon River. An independent community sprang up around Beardmore station, next door to the company townsite of Empire.

In the summer of 1934, a gold rush ensued along the Sturgeon (a.k.a Namewaminikan) River, a waterway north of Jellicoe flowing from east to west into Lake Nipigon. Rene Maloney's gold strike north of Twin Falls on the Sturgeon River sparked the staking rush. The river became a main transportation artery, even though the furthest reaches of the camp were only 10 to 12 miles from the railway. The railway crossed the Sturgeon five miles east of Jellicoe, and motorized freighter canoes could travel west from Sturgeon Bridge to Twin Falls in a single day, running the smaller rapids and portaging around falls.

The Sturgeon River mining camp seemed to fall naturally into two divisions, east and west, according to a major report issued by the Ontario Department of Mines. E.L. Bruce and H.C. Laird described the Sturgeon River Area in the ODM's 45th Annual Report, Part II, 1936.

The Western Division

The western division, closer to Lake Nipigon, comprised the townships of Pifher, Irwin, and Sandra, and parts of McComber, Summers, Meader, Eva, and Dorothea. Beardmore and Empire fell into Summers Twp.

Miners attempting to mount the Sturgeon R. from Lake Nipigon used smaller prospector canoes. They were advised to buy space on one of the fishing tugs based in Macdiarmid rather than attempt the rough waters and unprotected shoreline in canoes. Once in the river, they could use canoes.

An old lumber road called Sinclair's Road ran north from Beardmore to the south end of Long Rapids on the Sturgeon. Another old road, Poplar Lodge Road, ran from Beardmore toward the Standingstone River. Both roads were described as summer roads.

In the western division of the area, the terrain was quite rugged, with outcrops abundant, and greenstone (of volcanic origin) ridges trending in an east-west direction. A broad diabase ridge (Keweenawan period) traversed the volcanics, stretching from north of Long Rapids to the Blackwater River in the south. Sedimentary rocks underlay the flatter areas, which were often overlain with glacial and lacustrine deposits and muskegs.

The greenstone – highly metamorphosed lavas – was associated with other Keewatin series rocks, namely volcanic sediments (e.g., tuffs, agglomerates), flow breccias, and schists. Three almost parallel bands of sedimentary rocks, Timiskaming series, trended generally east-west in the southern part of this division. The rocks were primarily greywacké, associated with pebble or boulder conglomerate, slate, arkose, and banded iron formation.

Later intrusives (Algoma period) included granite, diorite, and porphyry, sometimes in large masses, sometimes in dikes.

The Sturgeon River was a deep, broad waterway. From Twin Falls to Lake Nipigon, the descent was rapid, falling about 200 feet in 11 miles. The main tributaries were Crooked Green Creek, joining from the north, and Foxear Creek, from the south.

At Poplar Lodge on Lake Nipigon lived some aboriginal families. On the railway in the western division, only Empire and Beardmore had significant communities.

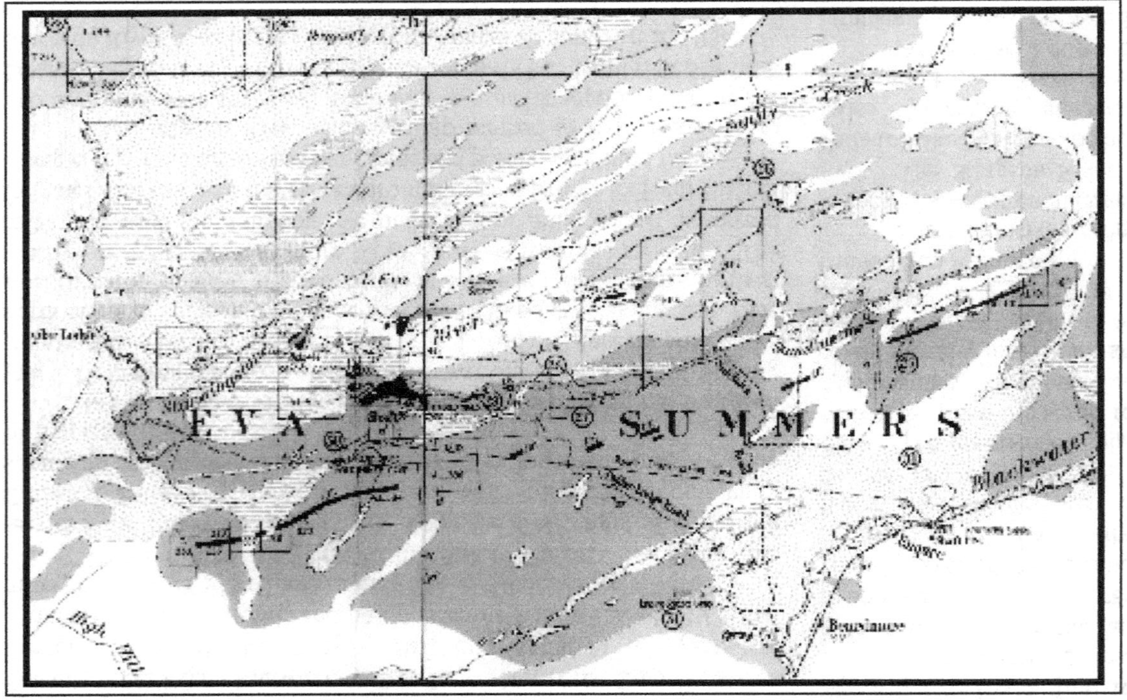

Section of Map No. 45a, Sturgeon River Gold Area, 1936. 26 = Calhet Exploration Syndicate. 27 = Long Lac Superior Syndicate. 28 = Wilport Gold Mines. 29 = Leitch Gold Mines. 30 = Sand River Gold Mining Co. 31 = Empire Contact Gold Syndicate.

A list of prominent properties in the western division in 1936:

Amorada Gold Mines, Limited (2 miles east of Lake Nipigon, 1 mile north of Sturgeon R.) - surface work (SW) only

Brennan and Kenty Bros Prospecting Company, Limited (west of Twin Falls, 2 ½ miles by trail) – SW

Calhet Exploration Syndicate (1 mile north of Standingstone Lake) – SW

Casey Contact Gold Mines, Limited, reorganized as **Brengold Mines, Limited** in early '36 (just south of Sturgeon R., winter road to Pasha L., then wagon road to Nezah on the CNR) - SW, diamond drilling (DD), 223-foot shaft

Connell Mining and Exploration Company, Limited – (north of Windigokan L., wagon road to Nezah) – SW, DD

Cyril Knight Prospecting Company, Limited ("East Group" at the mouth of Crooked Green Ck., "West Group" near the Brennan & Kenty Bros. property) – SW

P.E. Hopkins Group (just north of Twin Falls) – SW

K.L. Exploration Company, Limited (north of Windigokan L.) – SW

Lac-Teck Gold Mines, Limited (southwest of Windigokan L.) – SW

Leitch Gold Mines, Limited (south of Eva L. & Standingstone R., connected by wagon road and hydro line to Beardmore) – SW, DD, 350-foot shaft projected, a 65.5-oz. gold brick poured

Long Lac Superior Syndicate – (between Standingstone L. & Empire, adjoining the Wilport property on the south) – SW

Macjoe Sturgeon Gold Mines, Limited (just south of Sturgeon R., adjoining the Sturgeon River mine) – SW, DD

Maloney Sturgeon Gold Mines, Limited (main group 1 ½ miles north of Twin Falls, where a strike in 1934 precipitated the gold rush) – SW

Milroy-Taylor Group (north of Twin Falls, adjoining the Hopkins property on the east) – SW

Sand River Gold Mining Company, Limited (south of Standingstone R., between L. Nipigon and the Leitch mine) – SW, DD, 500-foot shaft projected

Karl Springer Exploration Company, Limited (north of Windigokan L.) – SW

Wilport Gold Mines, Limited (adjoining the Leitch mine on the west) – SW, DD

Laird's report listed several other properties but noted, "No important discoveries were made . . ."

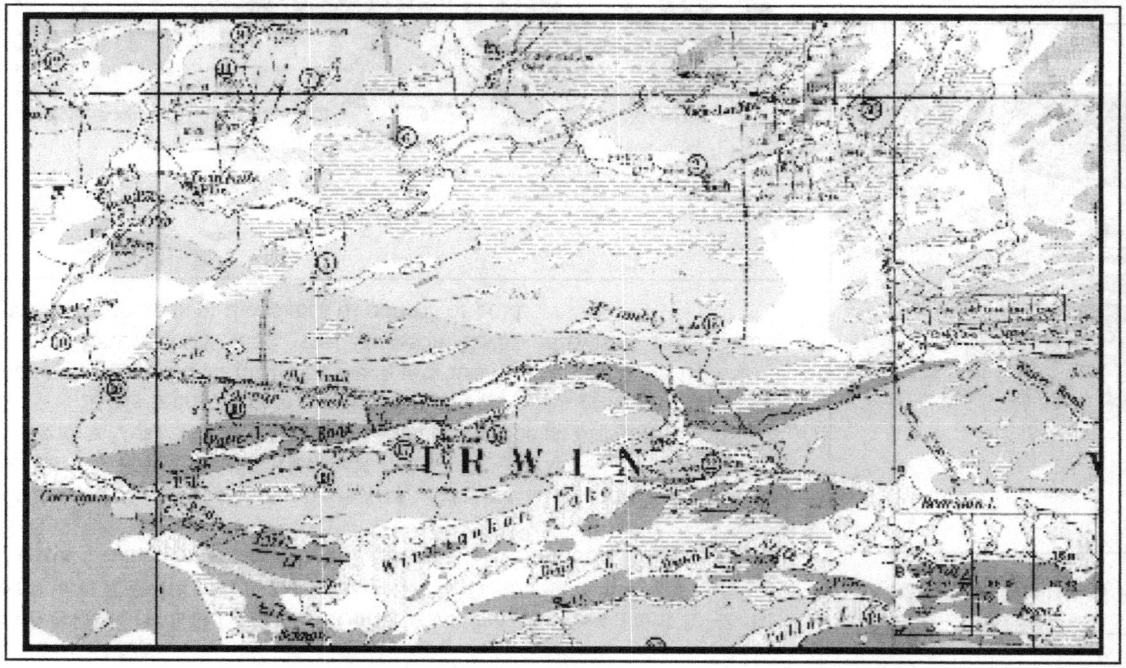

Section of Map No. 45a, Sturgeon River Gold Area, 1936. 1 = Macjoe Sturgeon Gold Mines. 2 = Brengold Mines. 3 = Minerals Development. 6 = Cyril Knight Prospecting Co. 7 = Milroy-Taylor Group. 9 = Taylor-Segsworth Claim. 11 = P.E. Hopkins Group. 12 = Spooner Sturgeon River Syndicate. 15 = Devanney-Smith Group. 16 = Connell Mining and Exploration Co. 17 = K.L. Exploration Co. 18 = Karl Springer Exploration Co. 19 = Minefinders. 22 = Ontario-Quebec Prospectors. 36 = Sturgeon River Gold Mines.

The Eastern Division

In the eastern division of the camp, staking and prospecting focused on the townships of Elmhirst, Walters, Rickaby, Leduc, Lapierre, Legault, Hipel, and Colter. Float (and ski) planes used Jellicoe year round as a base to service the country north of the railway.

By 1936 a couple of wagon roads ran north from the vicinity of the flagstop, Nezah, west of Jellicoe, in one case becoming a winter road as it traversed a large muskeg before it reached the promising prospects of Sturgeon River Gold Mines and Brengold Mines. Another wagon road ran west of Jellicoe to the Oremond mine. The Atigogama Road linked the Orphan mine, originally called the Dikdik, to the CNR at the Kinghorn flagstop on Partridge Lake, east of Jellicoe.

A thick mantle of glacial material created a fairly flat topography in the far eastern part of the mining camp so that it resembled the terrain of the Little Long Lac Area further east. The mantle thinned towards the west, and rock ridges created a more rugged terrain. Forest fires had destroyed practically all the timber between the railway and the river, leaving a tangle of charred logs and second-growth. North of the Sturgeon were stands of good-sized spruce, and on the ridges and sand plains, mature jack pine.

In the eastern division, sizeable lakes abounded, allowing fly-in access. The Sturgeon was the largest river, flowing generally west from Partridge L.; the Blackwater also flowed west from Jellicoe, generally along the line of the railway. There were three creeks which canoes could navigate: Just west of Seven Foot Falls on the Sturgeon, a creek drained the north country, providing a route to Onaman L., passing through Wilkinson L. Further east, a tiny creek drained Atigogama L. At the east end of Partridge L., Dumas Ck. entered, draining Dumas L.

Precambrian rocks formed the basement of the whole region, being lavas, tuffaceous rocks, and iron formation, many of which had been rendered into schists. Later, sedimentary rocks intermingled with the volcanics, being conglomerate, greywacké, slate, and iron formation.

Later than the volcanics and sediments came the intrusions of granite and diorite. North of Paint L., granite was prominent, and further north, past the Sturgeon, the underlying granite and granodiorite sloped gently upward.

A list of prominent properties in the eastern division in 1936:

Sturgeon River Gold Mines, Limited (south of the Sturgeon R., adjoining the Macjoe Sturgeon and the Brengold properties on the west, winter road plus wagon road link to Nezah, and hydro line to the CNR) – SW, DD, 500-foot shaft

Agaura Exploration, Limited (just east of the Sturgeon River mine) – SW

Orphan Mine (north side of Atigogama L., wagon road to the CNR at Kinghorn) – SW, an open cut, 250-foot shaft, production of 2,460 oz. gold in '34-'35

Oliver-Severn Gold Mines, Limited (midway between Paint L. & the Sturgeon R., with trail to the river) – SW, DD

Miron Claims (north of Coyle L., east of the Sturgeon River mine) – SW

Oremond Gold Mines (just west of Oxaline L., wagon road link to Jellicoe) – SW, DD, 150-foot shaft

Greenlee Gold Mines, Limited (1 ½ miles southeast of Nezah) – SW

In his report, Bruce alluded to "several other groups of claims", none of which had so far given much promise.

Before very long, the Sturgeon River Mining Camp would produce hundreds of thousands of ounces of gold.

Views of the old Northern Empire mine in the 1930's.

The about-to-be legendary Leitch mine gearing up in 1936.

Chapter 1 :
Year 2007

KODIAK EXPLORATION LIMITED

Activities of Kodiak Exploration Limited

Muskeg Tours 2009

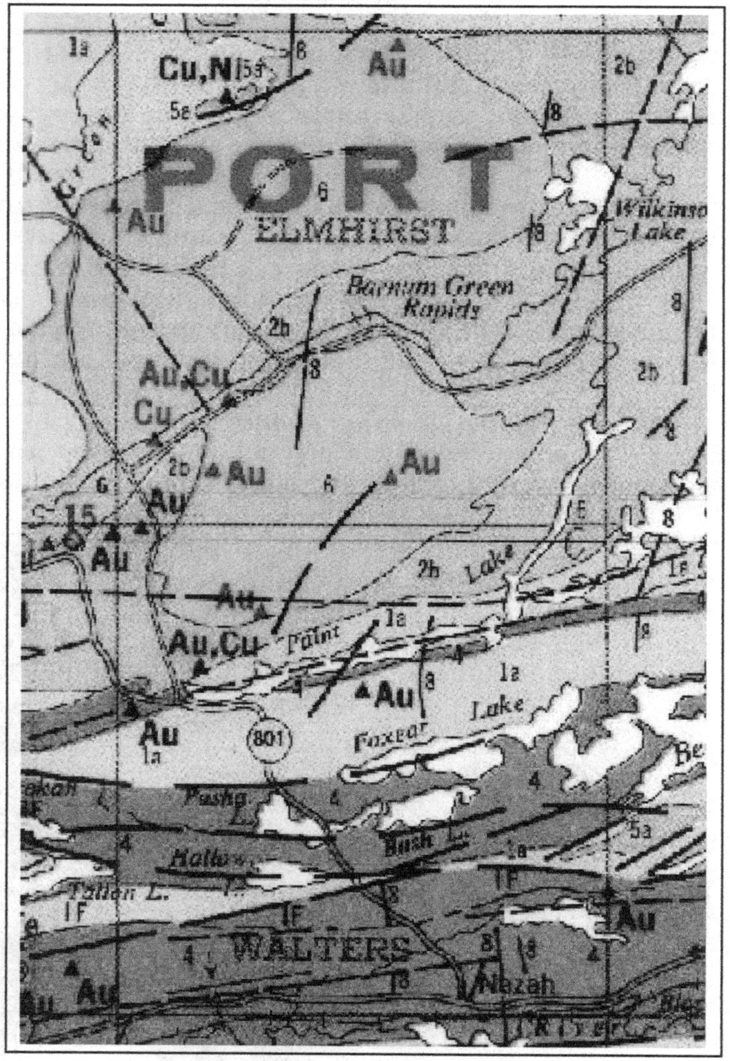

Old map with old roads

WILKINSON LAKE GOLD ZONE

What's it feel like to stand atop a potential gold mine? Good. Very Good. "VG," a prospector might say, although he would mean "Visible gold".

In August 2006, Kodiak Exploration Ltd. began stripping the overburden from a promising outcrop on the east side of the Kinghorn Road in ElmhirstTownship, north of Jellicoe. Oldtime prospectors had found this showing decades ago, but this was the first time that the occurrence would be extensively explored.

Workers took channel samples. They used a special power saw to cut out long slabs of rock a few centimeters wide. The samples were sent to an assay lab to determine the value of gold and associated metals.

The best value was 23.29 grams of gold per metric tonne over a length of 3.05 metres. A weight of 31.1 grams is a troy ounce, so 23.29 grams is more than .7, or 7/10ths, of an ounce. That's very good indeed. The gold came from a bonanza vein on surface. The market price for gold today is closing in on $670 an ounce in American dollars, or $700 Canadian.

On the recommendation of its on-site geologist, Stephen Roach, Kodiak began a drilling program. Through his research, he knew that others had drilled this zone before. "It has seen quite a bit of drilling in 1978 right to 1981," he told the Times Star, "and another spurt of drilling in the late '80's, accumulating around 5 kilometres of drilling."

On Monday, July 9, Roach led a party on a tour of the Hercules property of Kodiak. The party comprised Gerry White, District Geologist for the Ministry of Northern Development and Mines (MNDM); George Horobec, an interested citizen; and the Times Star. Observing the whitish dome or lens of rock for the first time, Horobec said he experienced a thrill. It had been a great many years since any citizen of the Geraldton-Beardmore Mining Camp -- who was not a prospector -- had been this close to a potential mine.

The dome appeared to be sitting in a puddle of water in a clearing. A backhoe had stripped away the surrounding bush and soil (overburden), and brought the vein-laced rock to the light of day. Every 15 to 20 metres, a shallow channel had been cut to intercept the main quartz vein at right angles.

The assay results triggered a 1500-metre diamond drilling program in late summer and fall over a length of 240 metres along the line of strike.

Stephen Roach, Gerry White, and George Horobec stand atop the original occurrence of the Wilkinson Lake Gold Zone. Behind them, the zone extends for about 200 metres to Wilkinson Lake.

The program involved 20 boreholes. "That [drilling]," said Roach, "intercepted significant gold results, up to 15.6 grams over 16 metres in one of the holes for these bonanza gold veins."

The official report on hole HR-06-03 indicates 15.59 grams of gold per tonne over an "interval" (apparent depth because of borehole's slant) of 16.6 metres and a "true width" (width of vein at its narrowest point) of 9.7 metres.

In explanation, Roach continued, "Bonanza gold veins are really high grade gold veins where you've got flour gold and coarse gold in the vein system. We have an envelope of low grade gold around these bonanza veins." Flour gold consists of very fine to micron-size particles that can best be seen with a hand lens or under a microscope. Coarse gold is VG.

With regard to the ongoing work on the Hercules property, Roach said, "Currently we're trying to expand the Wilkinson Lake Zone to the west, and find other multiple vein sets close [by], peripheral to the main structure here." The work has been occurring over a distance of 1.2 kilometres.

At this point the party walked westward to examine other zones.
[Series continued – Times Star]

Stephen Roach points out a channel cut in the dome. Behind him are stakes with flagging tape, indicating where holes have been drilled.

THE HERCULES PROJECT

Part 1

Walking west from the Wilkinson Lake Gold Zone, one crosses the Kinghorn Road and enters the bush. In a couple of minutes one encounters an enormous trench, 10 to 20 metres wide, that runs 350 metres, the length of $3^{1/2}$ football fields.

On Monday, July 9th, Stephen Roach, geologist for Kodiak Exploration Limited, led a party of three through the Hercules property. Gerry White, District Geologist for MNDM, was conducting an official field visit. George Horobec, observing as a citizen, and the writer, reporting for the Times Star, tagged along. The day was warm, with a light overcast, and few flies.

Stephen Roach, along with a local prospector, Amede Lafontaine, had rediscovered this strike in early August, 2006. (Refer to upcoming story *The Penelton Strike* for details.) By mid-August, a backhoe began exposing the quartz vein that is the heart of the Penelton Gold Zone (PGZ).

The trench began a few metres from the road and ran west about 90 metres before swinging northwest, the same compass bearing of the other zones the party was about to visit. A backhoe had stripped the vegetation and overburden (soil) from the bedrock, following a quartz vein less than half a metre wide. About 70 metres from the beginning, Roach pointed to a location underwater.

"We had 9.49 grams over 1.25 metres in here, and that was our best channel intercept," he said. He was referring to a channel cut at right angles to the line of strike.

Water filled all depressions in the trench. Most of it could be attributed to rainfall, and to the washing operation that removed the last vestiges of soil to expose the vein system. The overburden up to this point had varied from 1 to 3 metres deep. In the distance, the party could hear the noise of a powerful Wajax (a water pump) at work.

Just past the dogleg right, to the northwest, the vein "blew out". One of the challenges of the tour for the "tourists" was learning the lingo. A vein that widens dramatically is said "to blow out". One that pinches in, is said "to neck in". Other expressions for these phenomena are "pinch and swell", and "boudinage".

The quartz vein ran in the sheared contact zone between two rock types, the volcanic greenstone, which was the host rock, and the

Elmhirst Lake Intrusive, a granodiorite. When the vein blew out, it expanded to a maximum width of 9 metres.

The Penelton Zone begins by trending west. Depressions in the trench fill with water.

Very soon the party encountered a tiny cairn sitting on a knob of quartz. It served as a marker for GPS (global positioning system) readings. This was the point where Roach and Lafontaine had rediscovered the old Penelton strike. Gerry White remarked incidentally, "GPS has revolutionized the business when it really came in over the last 5 or 6 years."

Samples had been taken from channel cuts every 20 metres. Metal strips (tags) were fastened to the rock at strategic intervals along each cut, each with a number. Each interval is assayed separately, so that a 12-metre long channel cut, for instance, may have up to 20 samples for analysis.

The first samples a prospector takes from a showing are grab samples. With his hammer, he knocks off pieces of bedrock in a single location, and about two fistfuls constitute one sample. Each sample is assigned a consecutive number, such as 4412 or 4413, and a metal tag with the same number is attached with red flagging tape to a loose rock, and left where the grab sample was taken. The numbers are also recorded in the prospector's sample tag book.

Gerry White sits on the blowout (the point where the quartz vein swells dramatically). Below his hammer, note that the vein is only a few centimetres wide.

As the party proceeded along the trench, Roach remarked, "We're getting up to 10 grams on grabs here." He and White discussed the "anomalous numbers" they were getting in the Elmhirst L. Intrusion, but when they attempted to explain the term to the tourists, they failed.

At one location, Roach pointed out hematite that occurred as specks in granite. Other rocks nearby showed a dark brown rust called iron carbonate.

At the end of the trench, Roach read out the values in a grab sample at this point: 2.45 grams/tonne for gold, 16 g/t for silver, and 0.68% for copper. "Copper," said Roach, for the benefit of the tourists, "could be a pathfinder for gold . . . especially if you're prospecting." The more copper, the better gold values. In any future mine, copper would not be economic, but the silver would be recovered along with the gold.

Roach commented that three holes had been drilled in that zone last fall.

The party then followed a wide trail through low area for about 100 metres. It had been indifferently "cribbed" or corduroyed with tree trunks, over which the backhoe had traversed last fall.

The trench for the Yellow Brick Road Zone ran straight northwest for 310 metres. There had been one hole drilled last fall, yielding .41 g/t over 1.1 metres. One channel sample had yielded 2.6 g/t of gold over 1.51 metres, and 60.72 g/t of silver (also a good pathfinder). The best value had been .32 g/t for gold over 0.6 metre.

As in the PGZ, water filled all depressions. And as in the northern stretch of the PGZ, the persistent rainfall in spring and early summer had created seasonal streams that sometimes drained through the trenches and out through ditches.

Last year, suspecting that more showings would be uncovered to the northwest, Roach had sent Thunder Bay prospector Ray Koivisto to examine the ground. Koivisto found bedrock that had been dynamited by his forerunners. About halfway along the trench, the party inspected the location.

The party continued along the trench, examining the gold shoots in the granodiorite on the south or west side. According to Roach, at the north end the quartz vein "feathered out".

Part 2

A short walk westward over another wide trail brought the party to the Hydro Showing. Koivisto had noted overburden trenches in the area, and an area roughly 30 by 30 metres had been cleared. The trench yielded best values of 5.08 g/t gold and 27 g/t silver from grab samples.

Another short walk, and the Derek Lee Zone opened up left and right for about 100 metres. All the zones trended in the same northwest direction, and paralleled one another. Roach called them "multiple vein sets". The best grab sample was 2.39 g/t gold, 32 g/t silver, and 0.53% zinc. Roach consulted his geological maps.

"This is the most fascinating part of my job," he said, "the mapping." He carried a bound sheaf of hand-drawn maps that located and identified each rock type. The maps were creating a picture of the Hercules property that only a geologist could really appreciate.

The next trench to the west was named the 7 to 9 Zone. The name alluded to a character in the TV series *Star Trek Voyageur*, and to the circumstances of the zone's discovery by Koivisto. Roach explained, "Ray found it on Line 7W as well as on Line 8W. . . and found out a prospecting sample from 2006 was taken a few metres away from the quartz vein ending in [the numbers] 009." The zone also extends to Line 9W. The lines referred to the grid of trails that had been cut out of the bush to delineate claims. Stripping had begun this spring.

Referring to grab samples, Roach said, "We've got right up to 23.56 grams." At intervals of 15 to 30 metres along the vein, Roach had drawn lines in red fluorescent spray paint as guides for the workers who would cut the channels. Not far away, the Wajax, the noise of which had been growing louder, stopped.

An official report notes that one grab sample from this zone had an outstanding gold value of 102.20 g/t. An ounce is 31.1 grams.

About 150 metres from the 33 Road, the trench veered west, and stopped at the road. The party walked across the road and a few metres to the north. A loaded logging truck churned by, heading south. A clearing about 15 by 20 metres was the north end of the Marino Zone, cleared last fall, and named after a family in Geraldton. Koivisto had located this showing in a relatively flat outcrop that featured pits or "pot trenches", shallow holes blasted in the bedrock.

Mechanical stripping (with a backhoe) had begun three days before. A vein running southeast, only 20 centimetres wide, merged with a second vein, and then swelled to a width of 2 metres. The zone had been stripped for about 100 metres. The trench plunged steeply downhill and terminated at the edge of a swamp, and according to Roach, the vein system was still "open" to the south.

The two men who had been washing down the bedrock with a hose, had finished their lunch, and were about to make more channel cuts. Roach studied a slab of channel-cut rock with a lens, and then passed it around for inspection. Referring to the rock with its brilliant flashes of pyrite and chalcopyrite, Gerry White exclaimed, "That's what you call juicy-looking quartz!"

Everyone peered through a glass at a rock with a metal tag attached – a locator for a grab sample. Flour gold – visible gold – revealed itself as pinpricks of dull yellow.

One of the workers, Andrew Brown, born and raised in Geraldton, was soon starting his second year at Algonquin College. This was his summer job.

Mike Metansinine, from Jellicoe, was working his second year with Kodiak, having been laid off in the forest industry.

In his right hand Mike Metansinine holds a "locator" rock, speckled with gold, and in his left, a channel sample. Beyond him the Marino Zone trench stretches northwest.

Roach explained that there were four more zones to the north and west. The party got a lift back to the point where they had started their walk 3 hours earlier. Then, heading south on the Kinghorn, they made a brief stop at the Amede Zone trench, 130 metres long, located on the east side. Stripping had been accomplished in the last month, and was scheduled to begin again on the west side in a few days.

"This is the best target area I've seen going from this end to that," Roach offered, referring to the entire Hercules property. He pointed out that it has similarities to the Val D'Or Gold Camp in northern Quebec, that produced over 12 million ounces of gold. Alteration in the rock, and folding and fracturing and shearing, are good signposts for prospectors and geologists.

At the end of the week, on Friday, July 13th, the stock market listed the closing price of a Kodiak common share as 52 cents. The following Monday, a press release summed up the exploration and the results to date, available from www.kodiakexp.com.

On a return visit to the property on Tuesday, July 17, Roach pointed out to the writer a new vein system, recently discovered, that trended south from the north end of the Marino Zone. Roach thought that it extended for over 100 metres parallel to the 33 Road. In the course of discussion, he named and ticked off on his fingers several other showings that also merited more extensive exploration.

A drilling program is slated for August.

- Times Star

THE PENELTON STRIKE

The gold showing that sparked the exploration of the Penelton Gold Zone has a long history. Last year, on August 7, 2006, two men set out to extend that history.

It was a cool, rainy day, one that seemed to presage the onslaught of autumn. Starting at the Wilkinson Gold Zone, Stephen Roach and Amede Lafontaine took a compass bearing almost due west, at 250 degrees. They walked across the Kinghorn Road, at kilometre 32.7 from Highway 11, and struck off into the dense bush. Almost immediately they struck gold, so to speak.

Amede Lafontaine, prospector, and Stephen Roach, geologist, point out a drill casing in a long-abandoned overburden trench.

Roach was working for Kodiak Exploration Ltd. as a geologist. Lafontaine was a prospector from Beardmore who knew the country intimately, and was seeking to option his claims on this ground to Kodiak. Before they had taken their first steps, they consulted a map. It was a unique map, there being only one in existence.

It had come into the possession of Amede Lafontaine through his father, Endy, who in turn had acquired it from a prospector, Bill Garvey, about the time the Lafontaine family had moved to Beardmore in 1958. Measuring 43 by 58 centimetres (17 by 23 inches), it was drawn entirely by hand in ink, with some patches filled in with coloured pencil. A legend matched different colours to rock types. Most of the map was white, indicating "drift", that is, overburden, and the absence of rock outcrops. A few lead pencil notations seemed to indicate claim numbers. The scale was 3 inches to 20 chains, a chain being 66 feet.

Titled "STURGEON RIVER GOLD AREA – ELMHIRST TWP – ONTARIO", the map covered the the terrain between the northeast corner of Elmhirst Lake and the western reaches of Wilkinson Lake. The map showed no roads or trails, for there were none in those days. It did outline a block of sixteen claims. Here and there were two short parallel lines, the space between filled in bold red pencil. These indicated the location and trend of gold-bearing veins or showings. There were ten of them. The Wilkinson Lake showing occurred in the northeast corner. Just slightly south of west from this showing was another labeled "Penelton Claim".

On that rainy day in August, Roach and Lafontaine took a bearing between the two showings, and using the scale, determined the distance to be about 20 chains, i.e., 1320 feet, or 400 metres. Roach knew that 16 of his paces equaled 25 metres. Almost immediately they stumbled upon an overburden trench, a shallow pit in the soil with a diameter of a metre or so. Projecting from it was a rusty drill casing, evidence of historical work on the claim.

Still pacing off the distance, and following the bearing, the duo fought their way through the thickets. Back in the sixties, this area had been logged off by Domtar, and the brush and second-growth mixed wood made progress difficult. They came across other trenches, sides caving in, overgrown with vegetation.

A year later, on July 17, 2007, Roach and Lafontaine led the writer over that same terrain. The walking was easier, because a backhoe had created a trail that led directly to the original Penelton strike. Side trips into the bush revealed the occasional overburden trench and at one point the wheel ruts of a Timberjack from the logging era.

Arriving at the Penelton Gold Zone trench, stripped in late summer of 2006, the party stood on the quartz knob that had been rediscovered that rainy day. Lafontaine described the experience:

"Steve was facing ahead with the compass and I was tailing along behind, setting up some flagging tape to find this old showing. And when we got here, Steve was standing right here [indicating the high point of the exposure], and he said, 'Jeez, it's got to be around here somewhere'.

And I was standing just next to him here, and there's this little sapling here which had fallen over, and the roots were sticking out, and I could see some white. So I looked back again a little bit more, and there's a whole bunch of quartz, and so I said, 'Well, we're standing right on the showing'.

Well, Steve, right away, he took ahold of that sapling and just tore the whole thing off of the rock, and the more that he exposed, the more there was. And I had a hell of a time to get him off of there."

At that point, Roach and Lafontaine had a chuckle, and Roach said, "That's true".

The co-rediscoverers of the Penelton strike stand atop the quartz knob that was a year before disguised by trees, brush, and organic matter.

Lafontaine continued, "I had to stand back because some of the roots were hitting me in the face, and he's in there, just twirling [around], just . . . gone, just like a bear."

As Roach had peeled off the brush and moss, the quartz vein began to shape up. Addressing the writer, Roach said, "That was completely overgrown . . . What you see here is some brownish [weathering] discolouration? This was exposed to the atmosphere, or to the air, eh? So you could sort of see bits and pieces of exposure here. Where it was white, that's all new . . . "

The Bill Garvey map had some notations for the Penelton showing. One was "Vein exposed in trenche [sic] for 170 ft". Six chip samples, each taken across a width of six or seven inches, produced gold values ranging from $1.50 to 21.95 [per ton]. Those were the days when the price of gold was $35 an ounce.

Roach explained to the writer that within a day or two, he had asked Ray Koivisto, a prospector based in Thunder Bay, to do a thorough sampling of the showing. Koivisto peeled back more moss and stripped more trees and took several grab samples. "There weren't any high grade gold values in any of the samples," Roach explained, "but it indicates that there was gold in the system, and, you know, he exposed the structure a lot more. And from thereon in, we had the backhoe operator come on in and expose the better part of it along the length of it . . . We went thataway [gesturing southeast towards the Kinghorn Road, perhaps 130 metres away]."

In late summer of 2006, the Penelton Gold Zone was exposed for 350 metres following the northwest-southeast trend of the quartz vein, and Kodiak Explorations Litd. bored three drill holes. In the same period, several more gold zones were discovered on the northwest bearing [See previous story *The Hercules Project*].

Just to keep the record straight, be advised that the history of the Penelton strike really began in 1935. Refer to the upcoming story *The Wilkinson Lake Strike*.

- *Times Star*

EDITOR'S NOTE

Kodiak later renamed the Amede Zone, the Golden Mile.

THE WILKINSON LAKE STRIKE

At the southern end of the Amede Gold Zone, Amede Lafontaine kneels beside the spectacular quartz blowout on a property he has been exploring for decades. (Correct date is July 17.)

Part 1 – How It Started

In the old days, there were three ways a prospector could get around in the bush. He could walk, he could paddle, or he could fly.

So in late spring of 1958, when Amede Lafontaine wanted to visit the claims his family had acquired in Elmhirst Township, in the deep bush north of Jellicoe, he took his canoe, a 14-foot cedar strip. He launched in the Sturgeon River (Namewaminikan River) not far from the point where Wilkinson Creek entered. Amede paddled the stern, his father-in-law, Leon Michaud, the bow.

Wilkinson Creek (Today the maps call it O'Neil Creek) was in full flood, the water was swift. "... We'd take the odd little break here and there," Amede recalled, "but it was hard going. At one place we had to portage, there's a waterfall."

The canoeists ignored the portage from the creek into the south end of Wilkinson Lake because it entailed climbing up and down a steep hill. They proceeded to the point where the creek debouched from the east side. They turned north, passed through a narrows, and paddled across the upper pond of Wilkinson Lake to the northwest shore. Where a brook entered, they beached the canoe and walked a hundred metres or so up the right (south) bank till they came upon Bill Garvey's cabin. The cabin itself was in ruins. It was just a few metres further west overland to the Wilkinson Lake showing.

Endy Lafontaine, Amede's father, had recently acquired a hand-drawn map of this terrain from Bill Garvey, another prospector. Bill was interested in a property owned by Endy south of Timmins. "It was a trade-off", said Amede, and so the Lafontaine's came into possession of a map that showed the two major lakes, Wilkinson and Elmhirst, several rock outcrops, and 10 gold-bearing veins marked in bold red (See previously published story *The Penelton Strike* for more details).

On a table in his own kitchen, Amede Lafontaine peruses the sketch map drawn by Bill Garvey a long time ago. Amede's fingers rest on Wilkinson Lake, and following the direction his fingers point, one can see the "bar" that locates the historical showing which started his decades-long pursuit of gold in this area.

That was the first day Amede did work on the Wilkinson Lake strike. He would visit the property many times, always by canoe, over the next few years.

In 1934 the Sturgeon River area had experienced a gold rush. Hundreds of prospectors descended upon Jellicoe, a major divisional point, and Beardmore, a flag stop on the Canadian National Railway. No roads connected them to the outside world. As strikes were made and prospects developed, individuals and consortia punched trails and rough roads through the bush. Soon hundreds, perhaps thousands (No one knows the numbers), of men toiled away at promising prospects and developing mines.

The name Penelton first gained prominence in *The Northern Miner* newspaper dated August 2, 1934. "John Penelton has made an interesting find in the Jellicoe-Long Lac section on the Sturgeon River, according to reports from the field, and 12 claims have been staked for Minefinders Limited[,] upon which a shear zone of five to six feet in width has been traced about 1,500 feet."

The report mentioned ore shoots identified by trenching and crosscutting and panning, and "in some sections two veins of eight inches have been opened up".

Another report on October 18 stated that Minefinders "now have six groups of claims in the district". On January 2, 1936, another report referred to "the Penelton gold find at Square Lake". The find was located about seven miles north of the Sturgeon River.

Were all these reports alluding to the showings identified on the Bill Garvey map? Not likely, although we cannot know for certain. The "Penelton gold find", however, has a strong link to the map, where one showing is labeled "Penelton Claim".

A report on April 16, 1936, pinned down the location of the "Penelton gold find". "This Square Lake find," the newspaper stated, "was made in the fall of 1935 and resulted in the staking of a number of claims in the vicinity." A gentleman by the name of J.H.C. Waite, of Toronto, had taken an option on the property. The report continued, "The present diamond drilling campaign has been undertaken due to the almost impossible task of stripping the surface showing owing to the extremely heavy overburden." Part of the task would have involved transporting a backhoe by canoe.

There were resemblances to the Penelton Claim. "The surface work last fall was successful . . . in opening up a 75-ft. length with additional length extension indicated in a pit along the strike to the southeast." Sampling had provided an average grade of $4.20 (per ton) over an average width of 16 feet.

The Bill Garvey map, with a much later provenance, showed the Penelton Claim vein trending northwest-southeast, exposed for 170 feet, and with gold values between $1.50 and $11.55.

The J.H.C. Waite interest sank 4 drill holes on a northwest trend, all to shallow depths, at angles of 45 degrees. At the time of the report, assay results from the first drill hole resembled those obtained from surface samplings, which were described as indicative "of a wide low grade occurrence". The report noted "the occasional assay giving a value as high as $18". This analysis bore a strong resemblance to the initial analysis reached by Kodiak Exploration Ltd., the company currently carrying out exploration on the Penelton Gold Zone. A drilling program currently underway may change that analysis.

However, the last paragraph of the report proved most interesting. "Property location is one mile west of mileage 47 on the east boundary of the Nipigon Forest Reserve. The area is some 10 miles north of Nezah on the Canadian National Railway and approximately six miles northeast of the Sturgeon River Gold Mines."

Applying the directions and distances in the last sentence quoted above, as well as the previous one reference of "about seven miles north of Sturgeon River", one arrives at Elmhirst Township.

The reference to "the Nipigon Forest Reserve", however, could pin down the location of the "Penelton gold find" precisely. After an arduous investigation, the writer did just that. (See sidebar *Tracking Down Square Lake*.)

Part 2 – How It Turned Out

In his many trips to the Wilkinson Lake showing, Amede Lafontaine stripped the overburden by pick and shovel. Sometime in the '60's, when he was working for Domtar Forest Products, a road had been pushed to within two miles of the lake.

"One night," said Amede, "I was able to use the tractor, and brought it in to the showing and scraped it clean." Amede understated the challenge of traversing two miles of bush in the dead of night, exposing the vein, and then getting the tractor back to the worksite before dawn. At one point the machine bogged down in the creek that passed by Garvey's old cabin.

Recalling more details of that night, he said, "Actually, that night when the tractor was cleaning off the soil off the showing there, the main showing that I was working on, [the operator] managed to get a good chunk of rock off of there that weighed probably around 15, 20 pounds. It was full of mud so that I didn't even know what the heck I had but I took it with me."

Back at the worksite, he threw the chunk into his vehicle, and headed home, which meant Geraldton at that time. He stopped at the MacLeod-Cockshutt mine, which was the one district mine then operating, and asked them to process the rock. Then he went to his home on Beamish Avenue.

As he walked in the door, his wife, Shirley, said, "The manager [of the mine] wants to see you at the office right away." When he walked into the office, the girl at the desk said, "They're all upstairs waiting for you." Amede was still mystified.

As he walked into the room, he saw a group huddled by the big window and peering at his rock. One geologist said, "Have you had a good look at this sample?"

"No, I haven't, because I took this out at two o'clock this morning out of the bush, you know, on the tractor, and it was full of mud."

"Yeah," the geologist said, "but we washed it off. So you haven't seen it?"

"No."

"Come 'ere," he said. Amede walked to the window. "Look at all the gold in this goldarned thing."

The rock was peppered with gold. When he got the assay, it was 17 ounces to the ton. The mine took out a strip of gold and cut it in half, keeping half for their efforts. Lafontaine was handed a strip over 14 inches long.

"I kept the other half in my wallet for the longest, longest time, and I finally lost track of it."

Amede never lost faith in the property. He optioned it to different companies, and even those that did do a little work on it, walked away. At one point a company in Vancouver said, "If you stake

200 claims, we might come up and have a look at it." It was a ridiculous proposition. "As a prospector, you can't afford that," said Lafontaine. "...To keep [a claim] in good standing, you have to do $400 worth of work per year to be filed with the government . . ."

On July 15, 2007, Ray Koivisto, employed by Kodiak Exploration Limited as a prospector, carries out a bag of grab samples. Jim Leduchowski's backhoe is stripping the Amede Gold Zone west of the Kinghorn Road.

"Then I finally figures, well, maybe I'm not the guy to really advance this darn property." He had some showings further west of Wilkinson Lake, so he brought in Robert Cote to stake a block of claims for him. Cote was puzzled. Why was Amede giving up a perfectly good showing?

Amede explained. "The reason is very simple. I can't move the darn thing. Hopefully if I leave it open, somebody will come in, and if they can stake it and move it, then I'll have a chance to sell mine . . . I might not make as much but at least I'll make a little bit."

Official records show that Robert Cote staked TB 1187667 on April 3, 2002. The owners of record are Shirley Lafontaine and Geneva Nichols.

The "somebody" who came in was Stephen Roach, an exploration geologist. Looking over the country north of Jellicoe, he saw the potential. On October 14, 2004, he and two partners staked TB 3006416, encompassing the waters and shoreline of Wilkinson Lake, including the Wilkinson Lake showing beside the Kinghorn Road. Roach retained a 50% interest, and his partners, Pierre Maillet and Denis Laforest, 25% each.

On July 25, 2006, the TSX Venture Exchange approved of the option agreement that Kodiak Exploration Limited signed with Stephen Roach and his partners. At some point Stephen spoke to Amede, who, according to Stephen, was quite forthcoming about his strategy to lure a company to work his new claims. On August 7, Stephen and Amede made the important rediscovery of the Penelton claim (See previously published *The Penelton Strike*).

Subsequently, Kodiak optioned the Lafontaine property, including the current Amede Gold Zone and the Penelton Gold Zone, signing the agreement on August 31, 2006.

In August 2007 Kodiak completed the second phase of drilling on the Wilkinson Lake Gold Zone. The Amede Gold Zone now has been opened up for many hundreds of metres to the northwest. Kodiak's drilling program will wind up in the last week of September. Results will soon be released.

In less than an hour from Jellicoe by road, several thousand metres of gold-bearing veins now lie exposed to the light of a new day.

- Times Star

On July 17, 2007, Amede Lafontaine (on the left) examines a sample of rock with visible gold, uncovered in the Amede Gold Zone. One worker is washing down the muddy bedrock.

TRACKING DOWN SQUARE LAKE

Was there ever a place in the Greenstone region called Square Lake?

Square Lake is not a unique name; Ontario has several Square Lakes. But was Wilkinson Lake, in Elmhirst Township, north of Jellicoe, ever called Square Lake?

This oldtime prospector is outfitted for travel. He has a packboard and packsack, mosquito netting wrapped around his wide-brimmed hat, a rifle in his hand, and a rock hammer hanging from his cartridge belt. His dog is also burdened with a backpack and such accessories as a gold pan. Note the man's tie tucked neatly into his shirt between the buttons. MNDM Photo.

First, the writer perused ancient maps. One photocopied map of the gold rush era showed Wilkinson Lake as an unnamed lake. So when was Wilkinson named? E-mail inquiries to the provincial and federal geographic names agencies yielded some information. Their records indicated that the name Wilkinson Lake was officially entered on June 2, 1950, and that the lake was possibly named after a Provincial Land Surveyor. Before that date, was Wilkinson known as Square Lake? As for the place name of Elmhirst, the township's name was officially recorded on August 12, 1935. There was no information on Elmhirst Lake.

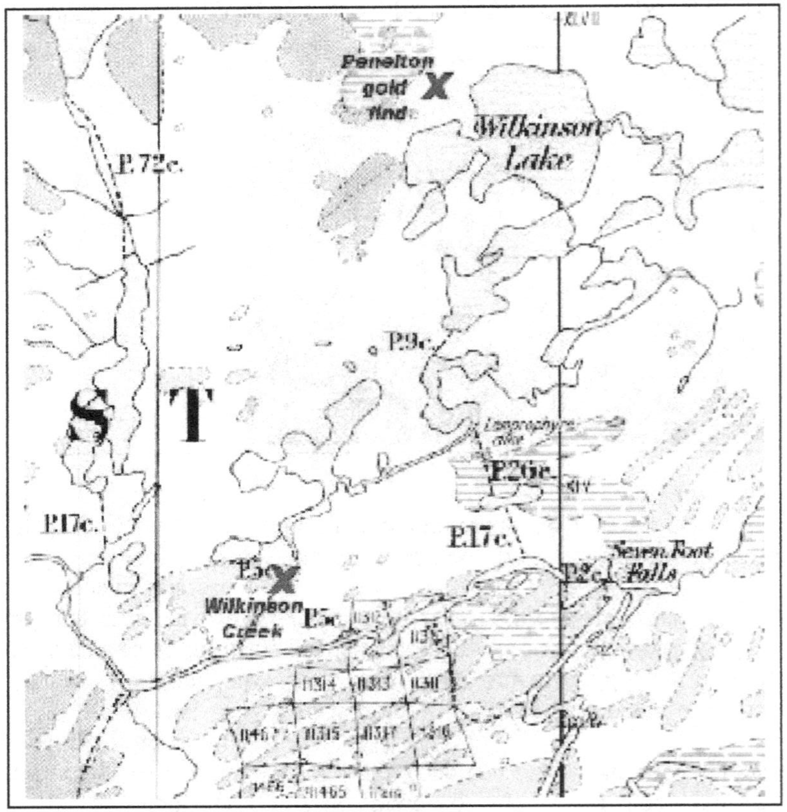

In this section of 'Sturgeon River Gold Area' (1936), the writer has inserted two "X"'s and associated labels. Note the squareness of the upper pond of Wilkinson Lake. Wilkinson Creek was later officially named O'Neil Creek. The township is named Elmhirst, but not the lake (below the "S" of the township name).

An important clue was the reference to Nipigon Forest Reserve. Searches on Google produced allusions to such a reserve currently located on islands in Lake Nipigon. The Ministry of Natural Resources is the agency managing the forested public lands of Ontario. Initial inquiries at the Geraldton Area Office and Nipigon District Office, proved fruitless. Officially, the Ministry does not have a long memory.

Was there ever a Nipigon Forest Reserve? Yes. The writer recalled coming across historical references to it, including some dating back to 1910. It was an era when no one was concerned about setting aside forest reserves in the unpeopled wilderness east of Lake Nipigon. Logging was, in fact, the order of the day, the wood being rafted down Lake Nipigon. Besides, Nipigon Forest Reserve seemed historically to be confined to an area south of Lake Nipigon, bordering the Nipigon River, a prime brook trout fishery, and a world-class attraction.

Pawing through old books and documents, the writer came upon a reference to the creation of a Nipigon-Onaman Game Preserve in 1933. Local MNR offices had no official memory of it. It covered 835 square miles, including Onaman Lake, north of Jellicoe. The writer recalled that in the early '70's, he had come upon a related sign tacked to a tree on a creek north of Geraldton. So the search took a different direction.

Years ago the writer had bought from an antique shop a full-colour laminated map of the Greenstone area in the 1930's. A little rummaging produced it – "Longlac", issued in 1938, a topographic map by Canada's Department of Mines and Resources. There it was – the Nipigon-Onaman Crown Game Preserve – at least the southeast corner of it. The southern boundary followed the CNR tracks from Beardmore, east to the CNR tracks between Longlac and Nakina, and followed those tracks north. No help. Those boundaries came nowhere near Elmhirst Township.

But Wilkinson Lake was labeled, and so was Elmhirst Lake. No Square Lake. But there was another piece of interesting information. The east boundary of Nipigon Provincial Forest ran north-south through Wilkinson Lake. Still no reference to Nipigon Forest Reserve. But the distance from the southern end of the boundary to Wilkinson was 47 miles, the distance specified in the old newspaper report.

A telephone call to the curator of the Nipigon Museum. Could they help? Betty Brill said they had recently received a batch of area maps from someone's estate in B.C. So the writer scheduled a trip to Nipigon. Dan, the summer student in charge, had received the heads-up. Judging from the stack of maps, it looked to be an all-day job, for there was no finding aid. Half an hour into the search, and bingo!

A beautiful colour map entitled "Sturgeon River Gold Area", which had accompanied the Ontario Department of Mines Annual Report of 1936. Wilkinson Lake had a name – Wilkinson Lake. But Elmhirst Lake was still unnamed.

The Sturgeon River Gold Area map of 1936 is an extraordinary document. A miniscule area in the vast province of Ontario is drawn and labeled in painstaking detail, half a century before GPS and satellite photography were commonplace tools. The map measures .55 by 1.1 metres (21.5 by 43.5 inches).

The map confirmed that the east boundary of the Nipigon Provincial Forest (not the Nipigon Forest Reserve) lay only a mile east of the historic Wilkinson Lake showing. The writer was ninety percent certain that the old "Penelton gold find" and the Wilkinson Lake showing were one and the same. By the time Bill Garvey created his map, the name of Penelton had been reassigned to a location further west in Kodiak's current Penelton Gold Zone.

Not entirely satisfied, the writer drove to Beardmore to attend the weekly prospectors' meeting. He explained his problem, and asked if anyone had documents from that era. Robert Cote said he did. At his home, Robert produced a marvelous map, issued in 1926. It was obviously a much-loved map, for it was falling to pieces.

The "Tashota-Onaman River Area" map, which had accompanied the annual report of the Ontario Department of Mines, showed the "East Boundary of Reserve" line running through an unnamed lake easily identified as Wilkinson. It coincided with the eastern boundary of the later Nipigon Provincial Forest.

Better yet. In big bold caps, between Lake Nipigon and the east boundary line, were the words NIPIGON FOREST RESERVE. The last proof!

To be fair, it was the last documentary proof. Earlier in the day, the writer had dropped into the home of Amede Lafontaine, and in the course of conversation, Amede remarked, "Square Lake? Of course I know Square Lake. I spoke to Bill Garvey many times about it." Square Lake was what the maps called Wilkinson Lake.

Who would have thought the name would persist in memory for decades after the lake had received its official name circa 1936? Even Garvey's map called the lake Wilkinson.

And if one looks with a critical eye at the contours of Wilkinson on a map, one perceives that the upper pond of that complicated water body does resemble a square, especially on the earlier maps.

Here's the epilogue. Dave Barker, MNR information management supervisor, working from the Geraldton office, had become intrigued by the writer's quest. He personally searched the District office, and enlisted the help of the MNR archives in Peterborough. There is, apparently, no official memory of a Nipigon Forest Reserve.

Still, the above exercise demonstrated that for prospectors and volunteer curators and amateur historians and for at least some government officials, history is alive and well in the North.

- Times Star

THE GOLDEN MILE

Part 1

Kodiak Exploration Limited has released the results on its newest gold zone on the Hercules property, and the values are spectacular.

On the morning of Tuesday, October 23, the writer toured the latest find accompanied by Steven Roach, Kodiak's project geologist, and Gerry White, district geologist for the Ministry of Northern Development and Mines.

The Golden Mile Zone, formerly named the Amede Gold Zone, begins near the Kinghorn Road (Kilometer 31.6) in Elmhirst Township, north of Jellicoe. The zone has been mostly stripped of overburden for a distance of almost 2 kilometres (Imagine 20 football fields end to end).

Taking the "33" Road off the Kinghorn, in 0.5 kilometres the party reached the access road leading to the Golden Mile Extension, passing a "No Hunting" sign tacked to a tree. The two pickups proceeded westward for exactly 1 kilometre over very rough ground. It was ground for vehicles with 4-wheel drive.

The writer had visited this point on August 25. Since then, the backhoe had skipped perhaps 40 metres of swampy ground, and then scooped out the trench again for perhaps 700 metres. A tree-topped hill in the distance marked the end of the stripping operation.

Stephen Roach led the way, walking around the swampy ground and descending into the trench. At this point, a spectacular grab sample had been scooped by Ray Koivisto, the on-site prospector, and a sharp discussion had apparently ensued about whether the sample should be bagged for assay or should be preserved as an exhibition piece. It ended up in the bag, and the assay produced the astounding result of 6000 grams per tonne (gpt).

Gerry White and Steven Roach discuss the splendid grab sample located at this spot in the Golden Mile. It assayed at 6000 grams per tonne.

At 31.1 grams per ounce, the sample assayed 193 ounces to the tonne (opt). Compare that to an ounce to the tonne, which is considered high grade ore.

Grab samples are, by their nature, very small samples. Better evidence lies in channel samples and drill core.

Roach called the 365-metre stretch between this point and the beginning of the Golden Mile Extension to the southeast, the "jewelry box". "We're getting gold in the wall rock, too," he said, referring to the Elmhirst Intrusive, a granodiorite. Channel samples had been taken every 15 to 20 metres. "In all these channels," he said, "we see VG [visible gold], and we didn't even need [magnifying] glasses." The channels averaged 20.20 gpt over an average width of 3.8 metres.

From the east came the sound of a backhoe filling in a cross trench. The Wilkinson Lake-Penelton-Yellow Brick Road-7 to 9-Marino line of strike had a bearing of 340 degrees. An exploratory trench had been dug between that vein system to the east, and the Golden Mile Zone, and in order to comply with regulations, its was being backfilled. Both Roach and White highly suspected, but had no proof yet, that there were more parallel vein systems between the two major exposures.

Steven Roach stands on one quartz vein that streams towards the camera. On the left side of the boulder, another vein joins it to form a vein 11 metres wide in the foreground. In the distance is the end of the current phase of stripping the Golden Mile.

At the end of September, Kodiak had completed the third phase of drilling, 36 holes. The holes were "shallow", most of them between 50 and 100 metres in length. The published results (See news releases on www.kodiakexp.com) included this statement: "Highlights . . . include drill hole HR07-29, which graded 54.1 gpt (1.57 opt) gold over a mineable width of 2.0 m, including 134.4 gpt (3.9 opt) gold over 0.8 m, and drill hole HR07-31, which graded 12.64 gpt gold over 1.4 m."

Both channels and cores confirm continuous gold mineralization on surface and at shallow depths. In the next few months, further drilling will explore at depth. The objective is to outline an ore body, i.e., enough ore to justify developing a mine.

In a news release dated October 1, Bill Chornobay, President of Kodiak, commented: "We may only be seeing the tip of the iceberg . . . . These are very early days, yet we are seeing high grade in almost every structure we uncover."

Part 2

On August 25, the writer had visited the Golden Mile Extension with Stephen Roach. Being an extension of the Amede Gold Zone to the southeast, its temporary name was the Amede Extension. Both areas are now the Golden Mile Zone.\

The pencil point indicates a good-sized bleb ("nugget") of gold in a character sample from the Golden Mile on August 25.

The primary purpose of the field visit was to investigate the drilling carried out by North Star Drilling. Roger Belzile, the driller, and Kenneth "Kenny" Campbell, the helper, were drilling Hole #10 (HR07-10) close to the beginning of the zone extension. About 150 metres of very wet ground, which had not been stripped, separated the extension from the Amede Zone proper.

As the writer prepared to move on, Roach said, "Do you want to see some gold?" The reponse: "Of course!"

Roach led the way into the trench and up to a large knob of quartz sprinkled with pennies. Yes, one-cent coins. A closer examination revealed that each penny marked the occurrence of a "bleb" of gold (Think of a bleb as a bubble or nugget). At this point the vein was 7.8 metres wide, and the quartz was laced with gold. It fairly took the breath away. "I've never seen anything like this," said Roach, "on a grass-roots exploration where you do find something."

Roach had been using a channel-cutting saw to remove what he called "character samples", rocks he would have cut and polished for exhibition purposes.

The writer was sworn to secrecy until Kodiak Exploration Ltd. published the find on October 1.

Proceeding that day along the trench to the northwest, the party visited the backhoe in operation. Jim Leduchowski, of Steve Leduchowski Trucking Ltd., was building a pad at the head of trench from which the bucket could reach down and grab more dirt. He was also damming the running water that threatened to fill the hole, for he had encountered very swampy ground.

Ray Koivisto, prospector, was in the hole examining the bedrock and clanking about with his hammer. He was determining the trend of the vein. The rock washing process had not reached this area yet, so dirt and mud covered most of the exposure.

Koivisto took a few minutes to chat. He explained that he had found a showing about a kilometre northwest of the Amede Zone. Recently he had been guiding the backhoe along an overgrown tote road towards that showing. When the road swerved close to the line of strike of the Amede Zone, he made a decision to test the ground to the left (west) of the road. The bucket turned up chunks of quartz. They were now working on the vein, and had exposed it for about 325 metres.

Almost two months later, on October 23, the party continued northwest on a bearing of 305 to 310 degrees, following the trench and leaving the "6000 gpt" showing behind. Gerry White was taking a lot of digital photos of the rock types and structures, and collecting the odd sample. The vein itself exhibited a remarkable ribbon texture, revealing multiple intrusions of quartz into the ancient rock fracture. Water had pooled in the low spots due to the sustained and unseasonably wet weather.

Steven Roach checks on the backhoe excavating at the north end of the Golden Mile Extension on August 25. Beyond is a short stretch of swamp that the backhoe will skip. In the trench Ray Koivisto examines the bedrock.

The party approached a mound on top of which perched a large granite bounder. A quartz vein streamed to the southeast on either side of the boulder and merged into one vein. At this point the vein widened to 11 metres. No channel samples had been taken beyond this point.

On the east side of the boulder, a showing (an exposure of mineralized bedrock) had become the starting point for the backhoe operator. He had then worked southeast until he encountered the short stretch of swamp.

The party followed the double vein system northwest through a jack pine plantation, the trees 2 to 3 metres high. A "No Hunting" sign greeted anyone coming up an old tote road from the west. The party walked up a gentle slope to the tree line and reached the end of the trench. The stand of intermediate aged jack pine on the north slope acted as a buffer between the exploration and an unnamed lake. Roach, referring to the vein, said, "It's still motoring!" Koivisto had also found a showing on the north side of the lake.

The party paused atop the original showing that had motivated Koivisto in August to suggest moving the backhoe northwest of the original Golden Mile (Amede) Zone. Moving a few steps to the east, the party encountered the second vein, exposed by stripping and running downhill to the southeast into the main trench. White snapped away, fascinated by the diorite chunks trapped in the vein system, and the very clean, almost "surgical", break between the diorite intrusion and the quartz.

The party descended the hill and returned to the "6000 gpt" showing. With permission, White took some samples to cut and polish for exhibition, while Roach went to give the backhoe operator new instructions.

Roach remained on site to do more geological mapping, an operation requiring great skill and meticulous detail. White described him to the writer as expert in his field. Roach had said earlier that he had completed 60% of the mapping of the Golden Mile, and would probably do another 20% before the weather closed him down.

The fourth phase of drilling begins in November, and the fifth, early in the new year, during which the Golden Mile will be tested along strike and at greater depth. The structure, as the saying goes, remains open along strike and at depth, meaning that the zone can be explored further to the northwest and the southeast, and at depth with longer drill holes.

In a news release dated October 22, Brian Maher, VP of Exploration, commented: "The proximity of this land package to road access, rail service, power, and a potential workforce distinguishes this land package and the Hercules project from other projects in Canada which are often located in high cost, remote areas."

It is interesting to see the TSX chart of Kodiak's stock prices, which spiked after the October 1 announcement of assay results. Stock that was trading around 50 cents in July closed on October 31 at $4.04.

No doubt many residents will dream tonight of a jewelry box nestled in Greenstone's Precambrian rock.

- Times Star

Kodiak has established its field office at the Resort – The garbage bin has to be secured

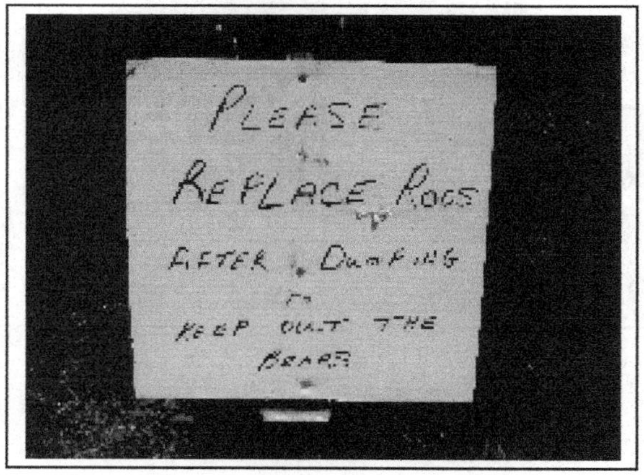

Chapter 2 : Year 2007

Activities of
Roxmark
Mines
Limited

Muskeg Tours 2009

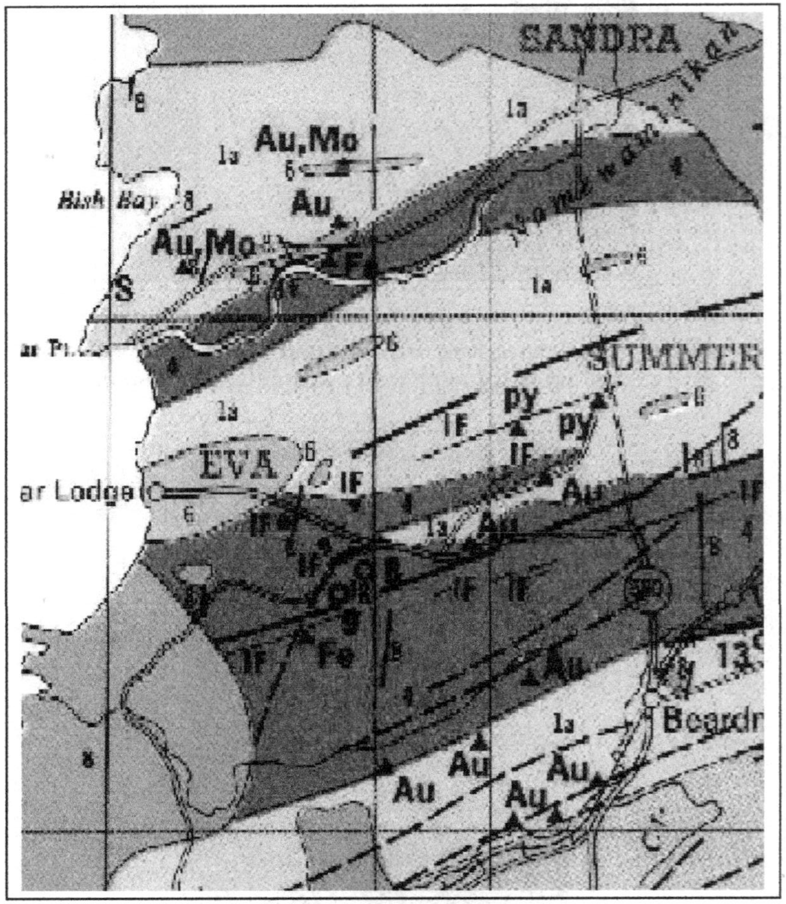

Old map with old roads

THE NORTOBA-TYSON PROPERTY

Gerry White, Glen Hart, and Louis Bernard stand at the east end of the trench. Note the rock bolts and wire mesh on the north face. The quartz vein behind them appears to be fragmenting.

"Watch your step now!"

Louis Bernard issued the warning several times in the course of the tour. He was addressing two experienced geologists and the writer, who knew his way around a keyboard but not a prospective mine.

On Monday, August 27, 2007, the party of four was exploring the Nortoba-Tyson property of Roxmark Mines Limited. Mining prospects are interesting places to visit, but they can also be dangerous, even to professionals. Bernard, Senior Vice President of Operations, usually directed operations in the Greenstone region from his Beardmore headquarters.

The north face of the trench rose straight up, and at its highest point towered 8 metres (25-odd feet) above the vulnerable humans below. Four-foot rock bolts driven into the face discouraged chunks of country rock from springing loose and squashing inquisitive tourists. Wire mesh overlapped the lip of the trench to secure rock debris from tumbling down. Even Glenn Hart, geologist for Roxmark, who had worked on the site during the excavation, and who was acutely conscious of the dangers, sometimes let his enthusiasm run away with him.

Gerry White, district geologist for the Ministry of Northern Development and Mines, was making one of his official visits. He hadn't seen the property since they were about to begin operations in August of 2006. The rubble associated with the operation still held many marvels and surprises.

Nortoba held a treasure, but it was not gold. It was molybdenum. The element had been captured by nature in the mineral molybdenite (pronounced Mull – IB – duh – NITE), often abbreviated to moly (MAWL – ee). Moly is a dark gray mineral that gleams and sparkles when it catches the light. It is a soft mineral that can stain one's hands. At Nortoba it occurs in thin films or layers, in flakes and nuggets, and in massive clumps. Moly, of late, is one of the

Glen Hart and Louis Bernard point out the safety fence surrounding the middle trench. In the distance, one can barely make out islands on Lake Nipigon.

most sought-after minerals in the world because of the upswing in steel production (See sidebar *Moly a Hot Item*).

Just before noon on that warm, lightly overcast day, the party had left Beardmore, taken Highway 580 westward for 4.5 km, and then turned north on a gravel road, the Camp 72 Road (well signed). The two pickups drove several kilometers through a forest of snags (standing dead timber) and green underbrush, scene of the great Nipigon 12 fire of 1999. Where the burn had veered northwest, they entered a mixed wood forest, and soon encountered a sign announcing "Tyrol Lake Demonstration Forest".

Back in 1949, some wise foresters had initiated a series of conifer plantations. The forest extended for several kilometers, continuing north of the Namewaminikan River. Just beyond the bridge (at the 11.5 km point from Hwy. 580), the party turned west on a single-lane road, and after a while emerged into the burn again.

The party stopped momentarily at a padlocked gate at the kilometer 18 point. Just over the rise, almost a kilometer ahead, one could spot a rocky ridge that had been assaulted by machinery. Skirting the site on the south side of the ridge, the pickups climbed a trail at a nearly impossible slope, and parked near the top of the ridge, above the system of trenches.

Dark snags, green brush, and low rounded hills dominated the landscape. Three kilometers to the west, hazy islands floated in the waters of Lake Nipigon. At the party's feet, three deep excavations ran on an east-west bearing for a strike length of just under a half kilometre (1,500 feet).

The new trenches (the "full cuts", as Bernard called them) followed a vein system that had been historically explored in twelve smaller trenches. The westernmost trench replaced historical trenches 1 & 2. The party walked eastward to survey the easternmost trench or cut, which had replaced historical trenches 11 & 12. Beyond it, Roxmark, when putting in the access road, had developed an extension of 24 metres (80 feet), and called it trench 13.

Flimsy net fences of fluorescent orange plastic outlined each full cut. Both Louis Bernard and Glenn Hart moved cautiously up to the fences, thereby setting good examples for Gerry White and the writer. Every so often Bernard and Hart picked up pieces of rubble and commented to the effect, "Here's a nice sample." The comment was usually followed by, "There's better to come."

The party descended the ridge a few metres between the middle and the easternmost trenches. One could observe the shallow pit trenches, numbers 8 to 11, blasted out of the bedrock by the early prospectors.

The middle trench stretched across the original trenches numbered 3 to 7 for about ninety metres (300 feet). At its deepest point, said Hart, the excavation reached about 13 metres (40 feet), but after the bulk sample has been removed, the trench had been partially backfilled with rubble. The trench opened at ground level to the west, but was still about 8 metres deep at the east end.

The party walked down the south lip of the trench and entered it from the west. White had pulled out a clear plastic sample bag and was stuffing it with pieces of rubble that held good moly. The writer soon followed suit.

Between August and November of 2006, Roxmark had removed a bulk sample of 2,000 tons from the property, and processed it at their Beardmore custom mill. Assay results have not yet been officially released.

Looking eastward up the trench, one perceived the north face (called the "mining face") rising straight up from the floor. The quartz vein system laced with moly had followed this face, and traces of it still remained. The south face (called the "rock face") fell back less steeply from the floor, with an occasional step. No rock bolts secured the country rock there.

Bernard pointed out that the quartz-molybdenite vein, called the No. 3 vein, had been staggered or offset along its length. The miners had followed it eastward for about 20 metres before it abruptly ended, and then had rediscovered it offset a few metres to the south. This occurred several times. The effect on the north or mining face, was a kind of "sawtooth" effect, as Bernard called it. At the end of the full cut the massive quartz vein had separated into a number of stringers, heading into historical trenches 8, 9, & 10. Traces of the vein still clung to the mining face. The vein's width averaged about half a metre (18 inches), but the depth ran from the floor to the top of the cut. Its depth under the floor was still unknown.

"I found visible gold in the moly," said geologist Hart. Hart pointed out that the early prospectors had been interested in this site for the gold. In fact, over the ridge to the north, there were three known gold veins, one sample of which had yielded an ounce to the ton – a very good value indeed.

At this location on the Nortoba-Tyson property, Roxmark plans to construct a portal to take operations underground.

The party exited the trench and stepped over a wall of hay bales, one bale high, positioned at the entrance to filter runoff. They climbed the steep trail and headed over to examine the westernmost cut. Noticing that the wind had flattened some of the safety fence, Bernard commented that he would have to come back to secure it. Roxmark officials frequently checked on that remote site. The locked gate and the safety fences signaled to the curious that this was a dangerous site for tourists.

The floor of the cut was covered by water to an unknown depth. The party hopped into the pickups and drove to the east end of the site, to a steeply sloping wall of native rock. Roxmark's plans called for an adit (a tunnel) in this face to carry on exploration underground. Bernard called it a "decline"; White, a "ramp". "What we'll do," Bernard said, " is essentially put in a sublevel drift that'll access the vein to [bulk] sample it."

In 2005, Roxmark had bored 24 holes on site, for 1,865 metres (1.2 miles) of core, but found the results spotty. The deepest intersection with the No. 3 vein occurred at 313 metres (1,028 feet). Judging by the erratic nature of the No. 3 vein, the results were understandable. The No. 3 vein had been traced on surface for 457 metres (1,500 feet) and by drilling for 717 metres (2,350 feet).

At the site of the proposed adit or portal, Roxmark had stockpiled the ore from its operations last fall. Bernard expected to see some rich moly samples still lying around, but apparently tourists had scavenged the area thoroughly.

Leaving the property, the Roxmark officials carefully locked the gate behind them. To date, there have been no reports of unauthorized visitors killed or injured at that site.

MOLY A HOT ITEM

As you read this, molybdenum is working in your body.

Molybdenum (pronounced muh – LIB – duh – NUM) is found in enzymes, with high concentrations found in your liver and kidneys, and lower amounts in your vertebrae.

Molybdenum is a hard, silvery metal, with the chemical symbol *Mo*, found in the periodic table between chromium and tungsten. It is one of the most useful commercial metals of the modern age. It was discovered and named in 1778 by a Swede in the primary source of that metal, namely, the mineral molybdenite, called moly (MAWL – ee) for short. Moly is a sulphide (MoS_2).

Commercial uses were slow in developing. The British used it in the Great War, armouring some tanks with 25 mm molybdenum plating. Since World War II, the applications for molybdenum have increased by leaps and bounds.

The single biggest market is its use in steel alloys and cast iron. Molybdenum enhances strength and toughness, improves resistance to corrosion (hence its use in stainless steel and tool steel), and makes steel easier to harden and to weld.

The metal can withstand extreme temperatures (melting point of 2610 degrees Centigrade) without expanding or softening. It is essential in such things as aircraft parts, electrical circuits, and heat exchangers.

It is becoming indispensable in the chemical, petroleum, and plastics industries. It's used in smoke suppressants (e.g., cable insulation), in lubricants, in pigments, and in agricultural fertilizers.

Molybdenum is selling in a bull market, lately hovering around $36 per pound Canadian (13 September 2007). World producers include the United States, Canada, Chile, Russia, and China.

Moly is mined by the open pit method, or by underground block caving, whereby large blocks of ore are undercut and allowed to collapse. This latter method describes how Roxmark Mines Limited will be taking its next bulk sample.

So now you know more about moly than you asked for. The next time you sit down to breakfast, remember that molybdenum is an important ingredient in your cereal, eggs, and toast. Enjoy.

Roxmark's custom mill processed some ore from the Nortoba-Tyson Property. Roxmark Photo.

THE NEW NORTHERN EMPIRE

Back in 1934, the Northern Empire mine, just outside of Beardmore, was the first operating gold mine in the Beardmore-Geraldton Greenstone Belt. Now Roxmark Mines Limited is aching to make it the first operating mine this century in Greenstone.

On the morning of August 27, 2007, Louis Bernard and Glenn Hart conducted a tour of the property for the writer and Gerry White, district geologist for the Ministry of Northern Development and Mines.

Louis M. Bernard, Senior Vice-President of Operations for Roxmark Mines Limited and a professional mining engineer, proudly showed the writer and White the custom mill on site, idled at that time, but capable of processing 200 tons per day of either gold or molydenite ore. Roxmark acquired the property in 1996, and has been earning revenue by processing gold ore for local mining explorers as well as ore from its own properties, which total 12 in the Beardmore-Geraldton area.

Roxmark's custom mill can process 200 tons a day. The process begins at the bin, upper left, where trucks dump the ore.

Glenn Hart and Louis Bernard pose by the ancient shaft. Beneath the concrete the shaft drops for 2,640 feet.

Glenn Hart, company geologist, pulled a core sample box from a rack near the mill. It was Box 26, NE-07-05, containing the core pulled from Hole 07-05 during the recent drilling program at the Northern Empire. The numbers 108.3 – 112.3 signified that the 4-metre length of core came from underground starting at 108.3 metres from the collar located at the surface.

Each drill core measured about 2 inches in diameter, a size technically called NQ. A tag in the box beside a split core (some short core lengths are sawed in half longtitudinally) identified Series F, No. 30302, from 108.3 to 108.87 metres, sent to "Swastika Laboratories" (SGS Lakefield Laboratories) to be assayed for gold. One split half went to the lab, one half remained in the box.

Hart licked the split core to enhance it it visually, and called out the minerals he could see with the naked eye: pyrite, pyrrhotite, and arsenopyrite. All those minerals, he said, were associated with gold at the Northern Empire as well as in the Beardmore-Geraldton Belt. Each tag had three identical parts: one remained in the box, one was sent away with the sample, and one was retained for office records.

The party of four then walked eastward a few metres to the mine's original shaft, now capped and crowned with an open metal framework (for a hoist). To the north, the terrain fell rapidly away to the now-disused CNR tracks, hidden behind the trees. Across the Blackwater River, the houses of Tansleyville popped up, and Empire Lake lay rather gray under the light overcast.

After climbing into two pickup trucks, the party bumped eastward along a newly opened trail for about 700 metres. The underground workings of the old mine never extended that far east. From May to July of this year, Roxmark conducted a 26-hole surface drilling program.

Historically, the old mine explored the Power Zone ore body. A parallel gold-bearing structure known historically as the Contact Zone was never exploited. Both zones trend in an east-west direction. The purpose of the drilling program was to confirm historical estimates of untouched ore, and to discover new bodies.

A 1999 report by a mining engineer, Peter Bevan, identified 69,969 tons grading at 0.47 ounces per ton, for a total of 30,789 ounces. The information in that report does not comply with a relatively new rule called NI 43-101 because the historical drill holes were spaced too far apart.

The Canadian Securities Administrators developed National Instrument 43-101 to govern how issuers disclose scientific and technical information about their mineral projects. CSA developed the rule after the Bre-X fiasco.

The old Northern Empire mine processed 426,000 tons of ore for a total extraction of 149,000 ounces.

The 2,640-foot shaft of the mine sank through a major diabase sill that lies parallel to the surface. All drilling this year was conducted above the diabase sill. Deeper exploration will necessitate dewatering the mine and drilling from points under the sill.

During the walk over the drilling site (from which most of the forest cover had been removed), Hart pointed out old drill holes that he had found, partly by following historical data and partly by good fortune. Some holes still had ancient casing projecting from them; others had been tiny openings disguised by vegetation and other debris.

Bernard pointed out that the small areas of exposed bedrock were greenstone. At the end of the trail, about 900 metres from the old shaft, in a newly bulldozed area, the party examined some bedrock that had not seen the light of day for many centuries. A diabase dike and banded metasedimentary rock had intruded the greenstone exposure from the south.

On three different dates – June 15, July 13, and August 13 – Roxmark released drilling reports for a total of 14 holes out of 26. In the August release, Monir Younan, President and CEO of Roxmark, said, "We are excited by these results. They will help to support an upcoming NI 43-101-compliant report which we believe will expand historical estimates and establish the viability of re-opening the Northern Empire Mine."

The same release reported, among other things, that "Hole 09-9B drilled on section 7+87.5E intersected 59.35 g/t of gold over 0.92 metre[s] (equivalent to 1.73 ounces per ton of gold)." He continued: "All holes drilled have intersected the Power and Contact zones, with mineralization persisting over a strike length of 212.3 metres."

To date there have been no further releases.

Some more exciting news transpired on July 12. Roxmark announced that it had optioned 61 claims covering about 1,580 hectares. The block, identified variously as "the Amede Properties" and "the Lafontaine properties", lies west of the mine and west of the community of Beardmore, on the line of strike of the Power and Contact Zones.

In the large image, a stick shoved down the collar identifies an old drill hole. In contrast, holes with casing (See inset) are easier to locate. Holes are usually bored at angles to the surface.

As the party was leaving the site, just before the gate, Hart drew everyone's attention to a promising showing beside the road, recently accidentally exposed by machinery. Although it lay 200 or so metres from the shaft, it had never been examined by the old miners. It was right on trend for the Amede Lafontaine properties.

Of its 12 properties in the Beardmore-Geraldton area, Roxmark is exploring only one for molybdenite. All the others Roxmark will target for its main quarry – good old-fashioned gold.

On Wednesday, September 26, the market price for gold hovered around $730., American and Canadian.

Spectacular folds in banded iron formation on the Buffalo Beardmore property

Chapter 3 : Year 2007

Muskeg Tours 2009

Activities of Premier Gold Mines Limited

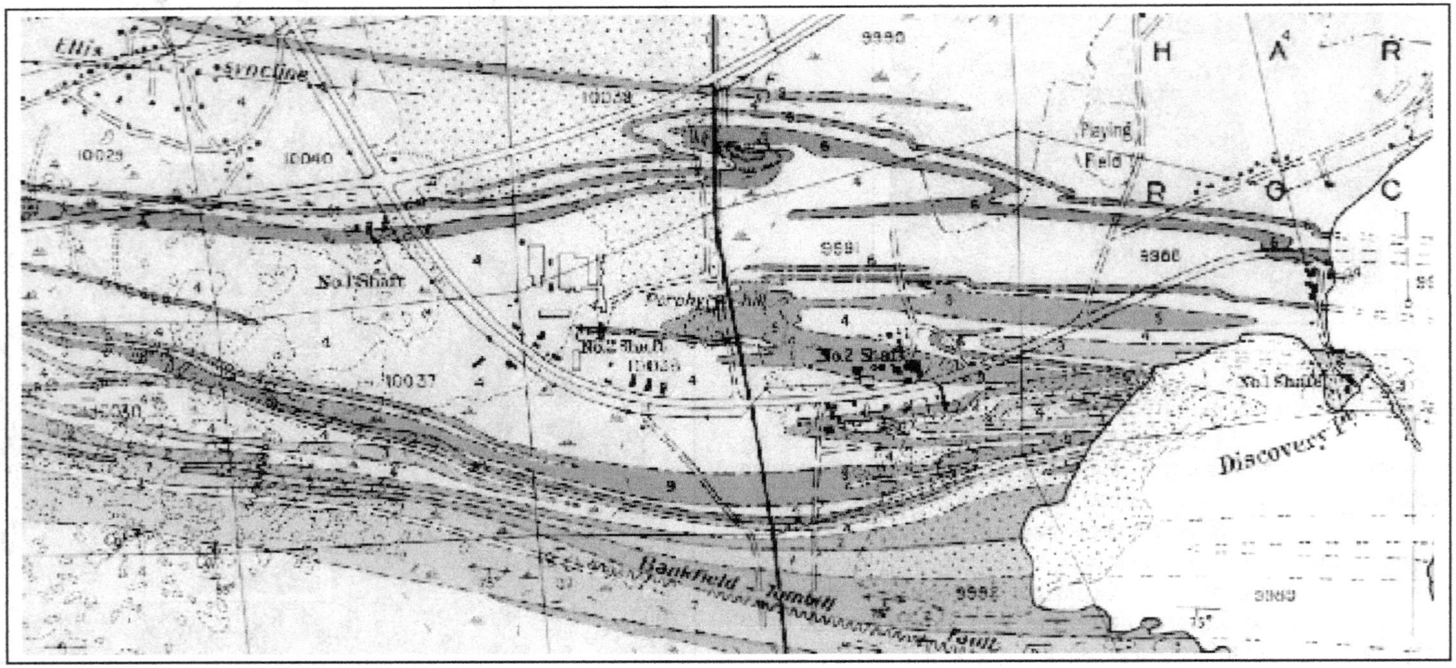

Section of Map No. 1951-2 "Township of Ashmore", ODM 1951

GOLD TIMES AGAIN IN GERALDTON

Last week many residents of Greenstone noticed the boxcar-like structure and tower set up on the "slimes" of Rosedale.

Premier Gold Mines Limited has begun a diamond drilling program on the mine tailings east of Hwy. 584 (Michael Power Boulevard), across from the site of the old Little Long Lac mine. Rosedale Point is the historic name of the little community that used to house the mine's staff, just south of the Barton Bay bridge. The Little Long Lac mine deposited its tailings east and south of Rosedale in the area that residents call the slimes.

Stephane Parent, President of Geraldton & District Chamber of Commerce, noticed increased activity in the retail sector to which he belongs. "We're very pleased that Premier Gold is shopping locally," he said.

Nancy Kyro, sales representative for Belluz Realty Inc., foresaw no immediate changes in the housing market, other than more activity for rental units. "At this time if you're going to sell your house, market values haven't yet changed." She said that prices could rise when Premier publishes the results of exploration sometime down the line.

There is a buzz of excitement among those who recognize that exploration is happening so close to home. Over the course of a year, Premier Gold will be investing $2.5 million in the vicinity.

Premier Gold Active in Greenstone Region

Friday morning, October 26, 2007, on the mine tailings of the old Little Long Lac mine in the community of Geraldton in North Central Ontario.

Chibougamau Diamond Drilling, based in Northern Quebec, has the contract. Their first drill rig arrived Tuesday, October 23. On Wednesday the crew were setting up at the collar of Hole No. 1, just a few metres from the highway.

Just after 10:30 on Thursday morning, the writer visited the rig, already under operation for 19 ½ hours. Shawn Vosdingh had just emptied the core barrel into the 3-compartment box, and put a little marker indicating that 51 metres of core had been pulled.

The NQ core (about 2-inch diameter) looked a rather washed-out green. Vosdingh, resident of Marathon, did not have the expertise to identify the rock.

On Friday morning, Chibougamau Diamond Drilling is operating two drill rigs in south Geraldton. They are set up on the slimes south of Barton Bay.

The driller, Martin Gagnon, of Quebec City, indicated that they were aiming for a length of 375 metres. The two-man crew had begun that day's 12-hour shift at 7:00 a.m., having also put in 4 hours the previous day after the set-up. At 7:00 p.m., another crew would take over. Crews were staying at the Crown & Anchor Motel.

Daniel Nolet, foreman, stated that the second rig had arrived yesterday, and would be set up and operating by Friday. He said that Chibougamau Drilling would be working in the vicinity for at least a year.

There was a flutter of concern earlier Wednesday when Public Works noticed the drill setting up. Rick Durocher, Public Works Working Superintendent, told the writer that Rob Cyr, equipment operator, had to confirm that the drilling would not be interfering with the municipal water line running to MacLeod Townsite.

A pumphouse on Barton Bay was supplying the rig with water through a 1 ½-inch line.

There had also been some concern from Rosedale Point residents about workers encroaching on their properties, for boundaries are not prominently marked. However, issues seem to have been resolved.

Premier Gold Mines has set up a permanent field office in the old air base building on Barton Bay in Rosedale. On Thursday morning, Glenn Harder, project geologist, was expecting to be set up with a telephone and Internet connection on Friday.

Glenn Harder, project geologist, exits the Premier Gold Mines office on Barton Bay, Rosedale.

"We're working in the vicinity of three old mines – Little Long Lac, Bankfield, and Magnet," he said, "and have other possible targets." This phase of drilling on the slimes will extend until Christmas. The Little Long Lac poured its first brick on December 17th, 1934, and the mine operated until 1953. "We're exploring the potential extension of the No. 9 Vein and the No. 10 Vein on the Little Long Lac," said Harder. "It's never been thoroughly explored." The company will be boring deep holes, up to 500 metres in length.

Glenn Harder has been hired to run the project for Premier. He had been working lately in Quebec and Newfoundland. Since 1980, he has explored in nine provinces. In 1981, he was staking in the Beardmore-Geraldton Gold Belt, and has been interested in the region's potential ever since. He is staying in the duplex near the office.

An October 24th news release by the company (www.premiergoldmines.com) stated that the project is a joint venture with Roxmark Mines Limited, and is named the Geraldton Project. Premier Gold can earn up to 70% interest in the property by making cash and share payments to Roxmark, and by performing exploration.

In a telephone interview, Stephen McGibbon, Executive Vice-President and Chief Operating Officer, said that the first phase will involve from 4,000 to 5,000 metres of drilling in the vicinity of Little Long Lac. Speaking from his office in Toronto, McGibbon said, "The initial holes were laid out by Ewan Downie, President and CEO."

In another brief telephone conversation, President Downie said, "We'll probably be looking at the iron ore potential too sometime in future."

Shawn Vosdingh and Martin Gagnon pose inside the rig, immersed in an unholy racket of machinery. On the left, the drill rod slants into the ground.

At 11:30 on Friday morning, two rigs were operating on the slimes. Hole No. 1 was pulling core up to the 162-metre point. A new helper, Joel Bedard, was assisting Martin Gagnon, driller.

At Hole No. 2, located farther east of the highway, the second rig was boring a hole through the overburden (tailings and soil). Dan Aube, driller, and Shawn Vosdingh, helper, had been working for an hour, and had gone down 12 metres using a "shoe bit", 3.75 inches in diameter, as opposed to a diamond bit.

Daniel Noel, foreman, said that Hole No. 1 required 21 metres of casing down to bedrock, drilling at an angle, and that the vertical distance was around 18 metres.

So, once again Geraldton is enjoying gold times. It is not only the oldtimers who remember Geraldton's heydays and who shiver a little in anticipation of prosperity.

PRESIDENT PROMOTES PREMIER GOLD

"We hope to have a big impact on Geraldton in the future." With this remark the keynote speaker introduced his slide presentation at the Geraldton District Chamber of Commerce annual meeting on November 8.

Ewan S. Downie, President of Premier Gold Mines Limited, was referring to the company's newest project, an exploration of old mining properties in the vicinity of Geraldton. On October 24, Premier Gold began a drilling program on the tailings of the old Little Long Lac mine on Barton Bay, Kenogamisis Lake.

President Downie pointed out that his company trades on the Toronto Stock Exchange, unlike most mining exploration companies, which trade on the Vancouver exchange. Premier Gold has its head office in Thunder Bay, another unusual feature.

Several local people have been hired at the project's field office, located in Rosedale Point on Barton Bay. The company, which started operations quite close to the homes there, has been trying to limit the noise factor by shutting drills down at night. "Diesel motors are not a quiet activity," he said, "and guys banging on pipes." When the rigs move further away, they will be operating 24 hours a day.

On the mine tailings, Premier Gold is targeting two gold vein sets. "The grade [of ore] of this mine was one of the better ones we find in Canada," he said. The company will probably bring in a third drill rig. Holes 1 through 5 had been completed, and Holes 6 & 7 were being sunk as he spoke. This phase of drilling requires 12 holes.

Addressing environmental concerns, Downie stated, "We will be reclaiming those areas where the drills were, and reseeding." Slides showed the disturbances to drilling sites.

Michel Lafrance presents a Times Star story on Premier Gold to Ewan Downie as a souvenir of his visit.

The 2007-08 program will involve line-cutting, an operation that controls the location of drill holes and assists in mapping the prospect. The winter program will include drilling through the ice on Barton Bay.

The winter program will also include drill targets on the old Magnet mine property, and on the "TAZ" property, as he called it. In a private communication, Downie wrote that TAZ refers to the Talmora Anticline Zone, located 2 kilometres west of Highway 584 (Michael Power Boulevard), and trending north-south.

In the early days, there was a mine located at the west end of Barton Bay, known as the Elmos , and in 1944, Talmora Longlac Gold Mines acquired the property.

"I view that [TAZ] as one of our most prospective targets . . . ," said Downie. "We feel Geraldton is a really underexplored area of the greenstone belt, and deserves more exploration."

Downie remarked that Algoma Steel has explored Barton Bay properties for iron ore. In 2008 Premier Gold will also test for a potential iron ore body. "Iron ore mines typically operate for decades," he said. He emphasized that a railway line is essential for a successful iron mine.

In his closing remarks, the President stated, "We're doing our best to minimize the effects on local people and at the same time using local people and business services."

In the question and answer session that followed, President Downie made the following responses.

"It will take a long time to prove up an iron ore body that's feasible," he said. In a private communication, Downie wrote that 6 to 8 years is the norm. At the meeting he noted that iron ore reserves in North America are depleting, and that iron ore is suddenly "attractive" today. A rail line would be essential to ship concentrated iron ore in the form of pellets.

The time line for a gold mine is less. "Four to six years would be a realistic time line for developing a [gold] mine in this region." Two years of drilling would be followed by environmental studies and a feasibility study.

The stock for Premier Gold Mines Limited closed on November 8 at $2.90 a share. Success at their Red Lake properties was driving the price, he said. Premier has four main projects: properties around Red Lake in the Red Lake Mining Division; properties around Pickle Lake in the Patricia Mining Division; a property in Mexico; and the Geraldton project.

As for economic grades of gold, he said that most mines in Canada operate on less than 0.2 ounces per tonne (6 g/t). "You probably need 0.3 or better in this region," he said (9 g/t). In its drilling program the company is looking for consistently good numbers.

The MC, Michel Lafrance, presented President Downie with a framed colour version of the Times Star front page on October 31, which featured a story on Premier Gold. He also presented a framed colour print, appropriated inscribed, of the drill rigs in that story, donated by Edgar Lavoie.

The company's website is www.premiergoldmines.com, and its stock exchange symbol is PG.

Beyond the rig is the subdivision of Rosedale

Chapter 4 :
Year 2007

SAGE GOLD INC.
Precious and Base Metals in the Americas

Activities of Sage Gold Inc.

Muskeg Tours 2009

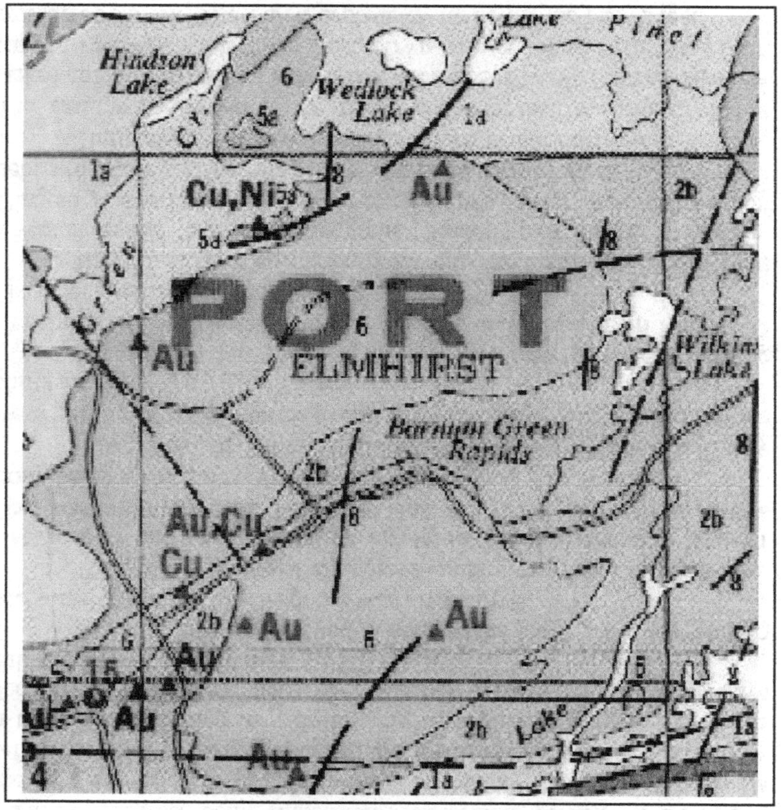

Old map with old roads

THE JACOBUS PROPERTY

Sage Gold Inc. has a property that could resemble a lot of old nickels.

Old nickels, as in five-cent pieces, used to be mixtures of the metals copper and nickel. One of Sage's properties, the Jacobus, is being explored as a copper/nickel prospect.

The Jacobus (often pronounced Ja – KOH – bus) is located north of Jellicoe and accessed by the 801 Road (Highway #801 and what was known historically as the Auden Road). Other Sage properties near Onaman Lake feature copper, gold, silver, zinc, and lead.

Dr. Ulrich Kretschmar points out the capped drill casings of holes SJ07-04 and SJ07-05. The caps kept out the weather and critters until geophysical instruments had been lowered down the holes.

On October 16, Dr. Ulrich Kretschmar, project geologist, guided the writer and Gerry White, district geologist for the Ministry of Northern Development & Mines. The objective was the historical trench with the copper/nickel (Cu/Ni) showing. The party left Bush Lake Resort in the late morning, where Sage has its field office, and proceeded north on the 801 Road, across the Namewaminikan River, to a point just short of the Kilometer 19 marker.

The Nissan Xterra turned right (east) on a rough road and bucked through several mudholes for 5.6 kilometres. Towards the end, a small diameter cable, blue in colour, ran alongside the road to the right. A field crew was conducting induced polarization geophysics (IP).

Further progress being impossible, Kretschmar parked the SUV, and the party started to hike. For a couple of minutes, the party clung to the left "shoulder", avoiding the quagmire that used to be a road. Recently, a tractor had skidded in a drilling rig to begin the third phase on the Jacobus. The unseasonably wet weather was keeping the waterways full and the water table high.

Kretschmar turned around to address the party. "Do you want to see some drilling?" The response was affirmative.

The project geologist cut across country through a mixed wood. The roar of machinery gradually increased. Lyle Reiner, of Thunder Bay, was pulling core for Box 90 of Hole #26 (SJ07-26), the rock at this point being a volcanic porphyry.

Eric Seguin, of Timmins, driller for North Star Drilling, said they were pulling about 60 metres every shift (12 hours), and conducting two shifts a day.

Sage Gold Inc. had conducted the first phase of drilling December 2006 (2 holes) and had begun a second phase in May 2007 (14 holes), which concluded in July.

Kretchmar now led the party towards the historical trench. First it was necessary to cross the quagmire churned up by the tractor and rig. Diligent searching and a quarter of an hour's labour produced enough debris to throw a makeshift bridge across the river of mud.

The party encountered casings which projected from drill holes. The casings descended to the bedrock not far below surface. Metal stakes and tags identified Holes SJ07-04 and SJ07-05. The casings were capped because, Kretschmar said, down-hole geophysics (IP) had been conducted earlier. The Cu/Ni deposit lay far below the party's feet.

A few minutes later the party traveled along a grid line (an arrow-straight path cut out of the bush) which soon intersected another grid line. A blue cable traced that line. At the intersection, two very rusty casings marked the location of historical holes. Flagging tape identified them as J8 (JCJ006) and J8 (JCJ005). Kretchmar said that this project had surveyed all historical holes for mapping purposes.

A few metres further the grid line emerged into a clearing at the top of a windrow of earth running at right angles to the line. Below the windrow lay the historical trench. Sometime in the past a backhoe had stripped the overburden and deposited it uphill of the trench.

Gerry White's left hand rests on an old pipe crammed down the casing of an historical hole. The pipe made the casing easier to spot in dense bush. White's right hand grasps the blue IP cable that is tracing one of the many grid lines on the property, cut specifically for the IP survey.

The party descended the ridge into the shallow trench, about 65 metres long by 6 metres side. On the far side, the southeast side, the hill fell away through a mature jack pine stand to an unnamed lake barely visible through the trees.

From the air, the historical showing must look like a rusty streak in the bush. Prospectors are drawn immediately to "rusty" rocks, for they indicate sulphides (minerals with sulphur). In this case the sulphides were pyrrhotite, chalcopyrite, and pentlandite. Pyrrohotite is an iron sulphide associated with pentlandite. Pentlandite is the principal ore of nickel. Chalcopyrite is copper pyrite, the principal ore of copper.

For many decades this streak of rust in the bush has been a holy grail in the quest for copper and nickel. Prospectors have dreamed about it; exploration companies have sunk packets of money into it. No one has yet defined an ore body on which a mine can be developed. Along the rusty exposure, pieces of broken rock glittered with the yellowish metallic sulphides.

The rusty bedrock dipped or sank to the northwest, under the grid line by which the party had approached the trench. The sulphide mineralization (a layer many metres thick) dipped under volcanics to the northwest.

On the other side of the exposure, to the southeast (lake side), pure gabbro rock met the mineralized zone, described by Gerry White as "mineralized gabbro", the mineralization being sulphides.

The exposure was pocked with shallow pits where prospectors had collected grab samples. Last year, Sage cut 144 channels about 50 centimetres apart in a sampling program. There was also a box-like pit from which Sage had extracted a bulk sample of 70 kilograms. "You want to get below the weathered material," said White, "to reveal the true character of the rock."

Leaving via a tote road from the western end of the trench, the party followed another blue IP cable for a distance. Ulrich Kretschmar explained that cables extended 1 kilometre on either side of a base line.

Driving down the access road, the party stopped at a white pickup truck owned by JVX Limited. Dean McNichols, from Espanola, sat in the truck, operating the transmitter for the 6-man IP crew. No one

Gerry White stands on the mineralized exposure of the historical trench with his hand on the volcanics. The sulphide-rich sill dips beneath the volcanics in a northwest direction at a 45-degree angle. Behind him, a flagged pole marks the pit (See inset) from which a 70-kg bulk sample was extracted.

---

else was around. Somewhere in the bush was Rob St. Michel, the crew chief and receiver operator.

McNichols explained that JVX was conducting 3-day "Spectral IP / Resistivity Surveys" on the Jacobus. They had one more day to go.

Back at Bush Lake Resort, Kretschmar laid on a delicious lunch for the writer and White. The owner of a couple of drill rigs had dropped by, hoping to land a contract. Exploration in Ontario has been so active this year that rigs are in high demand. Kretschmar hired him on the spot, and left for the bush to show him the job.

Kretschmar had indicated earlier the company's long-range plans. "It takes a year at least before you understand what's going on," he said. A bulk sample of 1,000 tonnes or more will be extracted for further testing. "We're looking at an open pit design to mine the upper part of the ore body that is dipping at 45 degrees." At depth, the mining would move underground.

Just as a matter of interest, really old nickels used to be comprised of 99.9% pure nickel. In 1981, the Canadian Mint switched to an amalgam of 75% copper and 25% nickel. Since 2000, the Mint has been issuing nickels in the proportions of 94.5% steel, 3.5% copper, and 2% nickel (in the plating only).

Still, copper and nickel are both hot commodities, and prospects in the Greenstone region are looking up.

RECENT HISTORY OF THE JACOBUS PROPERTY

The Jacobus (Ja – KOH – bus) property received its name from a mining exploration company.

Jacobus Mining Corporation Limited started exploration in 1957. Two other out-of-region companies subsequently investigated before Sage acquired Jacobus.

The Jacobus property consists of 24 claim blocks totalling 230 claim units (each being 16 ha) in the northwest corner of Elmhirst Township, north of Jellicoe, accessed by the 801 Road.

In May of 2006, Sage Gold Inc. signed an option agreement with N.W.T. Copper Mines Limited (a local company) to earn a 100% interest in the Jacobus.

In the summer, Sage completed a channel sampling program on the historical trench, which is an exposure 65 metres long and 6 metres wide. The average grade of the 144 samples was 0.64% copper (Cu) and 0.28% nickel (Ni).

Sage also extracted a bulk sample of 70 kilograms for metallurgical testing. Results have not been published.

Sage Gold Inc. conducted the first phase of drilling in December of 2006. The two holes intersected 6.3 metres, true width (the same in both cases), with values of 0.85% Cu and 0.54% Ni, and of 0.41% Cup and 0.38% Ni, indicating medium grade values. These confirmed the grades from historical drilling which had outlined 954,804 tons grading 0.43% Cu and 0.40% Ni. The first phase had extended the known resource a further 300 metres "down-dip" from the original 150 metres. Down-dip signifies that the distance was measured along the length of the mineralization.

The Jacobus deposit, a sulphide-rich layer or sill-like structure, dips below the surface at a 45-degree angle from the horizontal. The dramatic exposure occurs in the historical trench, after which the sill dips underground.

In addition, the sill also "plunges", meaning that the full length of it is tilted at an angle from the horizontal. Imagine taking a flat chocolate bar. Place it on a tabletop. Then lift the end closer to you so that the bar is angled at 45 degrees. The bar has a 45-degree dip. Then, holding it in the same position, tilt the bar by lifting the edge to your right. The left edge now slants towards the tabletop. Aim for a tilt or "plunge" of 40 degrees.

The dip of the Jacobus deposit is 45 degrees and the plunge is 40 degrees. The sill dips to the northwest at 45 degrees and plunges westward at 40 degrees east. The northwest trend (direction) of the sill is N60W, which translates into a compass bearing of 300 degrees. The sill plunges (tilts) to the west at an angle of 40 degrees.

In February 2007, a geophysical consulting firm conducted a "down-hole" IP survey in the two drilled holes, "confirming the continuation of a mineralized zone found in each hole".

Dr. Ulrich Kretschmar, left, checks with Lyle Reiner and Eric Seguin of North Star Drilling. Note the skids on the rig, and the water running downhill, contributing to the quagmire that resembles a road.

On May 9, Sage commenced the second drilling phase to further outline the mineralized deposit and to help make the historical resource (958,804 tons) compliant with NI 43-101[(1)]. Over the summer the company released results as assays came back from the core. A news report on July 24 began with "Sage has encountered additional high grade mineralization over significant widths at the Jacobus Nickel (Ni)-Copper (Cu) property in North Central Ontalrio":

Hole 11 (SJ07-11) : 2.12% Ni and 1.02% Cu over 6.6 m

Hole 12 : 0.93% Ni and 0.62% Cu over 7.3 m

Hole 13 : 0.91% Ni and 0.79% Cu over 10.3 m

President Nigel Lees stated, "We have identified several high grade sections which occur within very broad sections of mineralization and hence would represent excellent mining widths." He said the zone was traced down-dip for 580 metres and was "open laterally and at depth". One hole (#7) missed the sill but intersected high grade gold, "a pleasant surprise".

A company news release tabulated the results of 14 holes, which, together with the 2 holes drilled in December, totalled 4,440 metres.

In early October, Sage commenced a third phase of drilling to total 6,000 metres. "Two drills have been deployed to test the lateral and down dip Ni/Cu mineralization[,] confirm the historical resources and test the gold zone that was encountered in the previous drill program." (October 4, 2007 - www.sagegoldinc.com) The second drill did not actually arrive until later in the month.

A consultant's report in 1978 described the mineralogy: "The sulphide mineralogy consists of pyrrhotite, chalcopyrite, and pentlandite, which constitute between 4 and 6 percent of the gabbro within the mineralized zone." Of course, this report, and others, have described the occurrence in more detail, most of it incomprehensible to the lay person.

Suffice to know that the Jacobus is a nickel/copper occurrence. Pyrrhotite is an iron sulphide (FeS), often found with pentlandite. Chalcopyrite, also known as copper pyrite, is a copper iron sulphide ($CuFeS_2$), and it is the common copper ore. Pentlandite is the principal ore of nickel. It is an iron nickel sulphite with the complicated formula of $(Fe,Ni)_9S_8$. It occurs in pyrrhotite.

Again, perhaps it is enough to know that the Jacobus is a promising Ni/Cu occurrence.

Endnote 1 - The Canadian Securities Administrators developed National Instrument 43-101 to govern how issuers disclose scientific and technical information about their mineral projects. CSA developed the rule after the Bre-X fiasco.

KEEPING TRACK OF CORE

The core that is pulled from drill holes on a single property during mining exploration, if laid end to end, can sometimes extend to several kilometers over the course of a year.

Sage Gold Inc. is using a supplementary technique to keep track of core. The company is photographing the core.

On October 16, the writer visited Sage's field office, located at Bush Lake Resort on the 801 Road, northwest of Jellicoe. Dr. Ulrich Kretchmar, project geologist, said, "We photograph every millimeter of the core". Also present was Gerry White of MNDM.

On a computer monitor, Kretchmar ran quickly through the digital images of core from Hole #7, (SJ07-07). "S" signified Sage; "J", Jacobus; the first "07" was the year; the second, the hole number. Beginning with Box 1, he sized up the images and described the rock as "uniform feldspar porphyritic volcanics". Box 21 had quartz carbonate veining in altered volcanics with high grade gold (the fortuitous strike that will be explored in the third drilling phase). Box 91 held gabbro. Box 126 came from the bottom of the 539-metre hole.

The project geologist performed the scan in a matter of minutes. His calls came from his experience and his knowledge of the geology on the Jacobus Cu/Ni deposit. The primary technique to keep track of core is called "logging"; it is performed by a geologist, and it is labour-intensive. The supplementary technique of making a photographic record of core is catching on. Said White, MNDM geologist, "More and more companies are doing that now."

Later Kretchmar explained that digital photos also make the core accessible online by, for example, geologists in the Toronto head office.

In the yard, Sage had dozens of core racks lined up, each about chest high. Imexko Contracting, a local company, had manufactured them. To demonstrate the recording process, Kretchmarr opened three core boxes and placed them in a flat position upon a row of core racks. He climbed up and, from a standing position, aimed the camera lens at the core in each three-compartment box, taking three overlapping shots. He used a digital camera, a high-tech Nikon D200 with a 17-200 mm zoom lens.

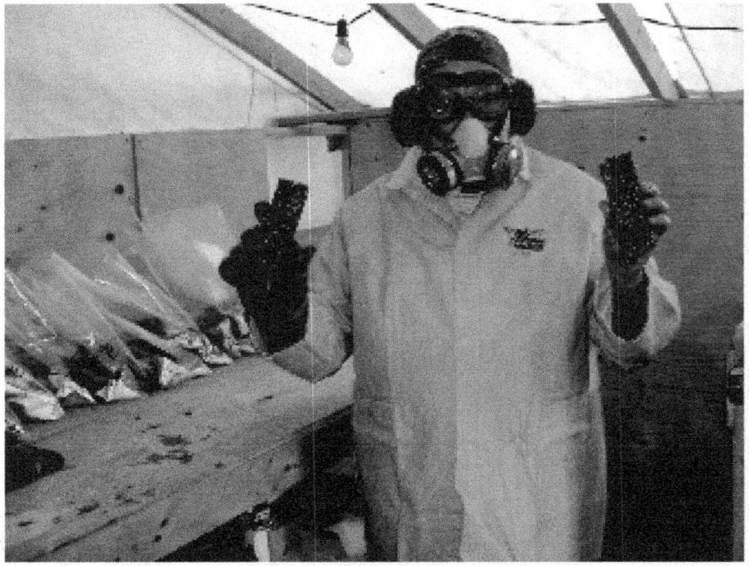

Inside the half-wall tent, Harry Saarinen, core cutter, holds up the products of his work. Behind him are sample bags which will be sent to the lab for assay.

In the vicinity of the racks was a half-wall tent with a plank floor. Harry Saarinen was splitting core. Suited up, ear-muffed, and masked, he looked a little like Darth Vader. The core-splitting process created not only a terrific racket but a lot of dust. A spray of water turned the dust immediately into mud.

Saarinen said he lived at Mountain Bay, 56 kilometres east of Nipigon on Lake Superior. He pushed a round core longtitudinally into a circular rock saw which was shrouded appropriately to contain the muddy run-off. One split half was deposited into a transparent plastic sample bag to be sent away for assay at ALS Chemex in Thunder Bay. The other half he returned to the core box.

At that point, Sage Gold had over 4.4 kilometers of NQ core from the first two drilling phases on the Jacobus, all pulled since last December. Another drilling phase had begun in early October.

Imagine keeping track of several kilometers of core. Most of us have trouble keeping track of monthly bills, which may measure less than 4.4 centimetres in depth.

Dr. Ulrich Kretschmar demonstrates how he makes a digital record of drill core. Gerry White stands by observing. The inset shows the three-compartment Box 91 from Hole #7 (SJ07-07).

Chapter 5 : Year 2007

Muskeg Tours 2009

Serving the Greenstone Region

Activities of Geraldton & District Chamber of Commerce

An interaction at the Chamber's AGM on November 8, 2007.

CHAMBER TO SUPPORT MINING EXPLORATION

"The main focus [this year] will be getting businesses ready for the mining exploration boom," said the incoming President of Geraldton District Chamber of Commerce.

Peter Kathmann addressed the crowd of almost 50 people at the annual general meeting on Thursday, November 8, 2007.

The evening began with a social at 6:00 in the French Club, followed by a buffet-style dinner at 7:00. Mike Lafrance acted as master of ceremonies (MC) for the occasion.

Some of the cheerful faces at the Chamber's annual meeting in the French Club on November 8.

The keynote speaker was Ewan S. Downie, CEO, President, and Director of Premier Gold Mines Limited. Premier Gold recently initiated a drilling program on the tailings of the old Little Long Lac mine near Barton Bay. Downie elaborated the company's plans to spend $2.5 million in the vicinity of Geraldton over the next year.

The next speaker, Jay Lucas, administrator of Greenstone Community Portal, focused on the new navigation strategies and the new look of the community website. (See upcoming stories on the presentations by Downie and Lucas.)

Larry Joy, executive assistant to The Honourable Michael Gravelle, Minister of Northern Development and Mines, spoke on behalf of the MPP for Thunder Bay-Superior North.

Other members of the head table included Sandra Rickard, consulting geologist; Lance Dyle, financial analyst; and John Begeman, director – all representing Premier Gold.

The MC, Lafrance, drew tickets for door prizes. Karen MacDougall won a laptop carrier embossed with the Canada Brokerlink logo. Jean Mason won a Premier Gold jacket.

For the business portion of the meeting, Jim Turner acted as Chair. First, Stephane Parent, the outgoing President, made some remarks.

Parent spoke about the many achievements of the Chamber in the past year. In the commercial sector he pointed to a successful Trade Show, and the coupon booklet initiative. In the civic sector, he referred to the region's first Clean-up Day campaign, launched on Saturday, May 6. Two hundred bags of garbage were collected. He commended Katherine Alton, the Chamber's representative to Northwestern Ontario Associated Chambers of Commerce. Alton was elected Vice-President of NOACC. Geraldton will be hosting the NOACC annual general meeting in the spring of '08.

Parent suggested three major goals for the Chamber in the forthcoming year. One, to give a positive response to the current mining exploration activity in the region. Two, to investigate the creation of a region-wide Chamber of Commerce. Three, to develop a five-year strategic plan for the Chamber.

Dave Barker, Treasurer, then presented two financial statements, one for the year ending December 31, 2006, and one updating the financials to November 7, 2007. It would appear that the Chamber is in a healthy financial position.

The Chair, Turner, presented a slate of candidates for the election of officers. Karen MacDougall accepted the nomination for the vacant position of Vice-President. There being no further nominations, the slate was adopted by the members.

Besides the V.P., officers will be President, Peter Kathmann; Executive Secretary, Karen Donio; Treasurer, Dave Barker; and Directors Karen Donio, Johanna Liscomb, Bill Mascotto, Huguette Venne, Ken Stevens, Jay Lucas, and Stephane Parent, also Immediate Past President. One director's position remained to be filled.

Katherine Alton will remain the liaison to NOACC, and act as the representative for the Municipality of Greenstone. Kathy Beaupre will be recording secretary.

The newly elected President of the Chamber, Peter Kathmann.

New President Peter Kathmann then addressed remarks to the meeting. He suggested that businesses must support the workers associated with the region's mining exploration boom. Restaurants, he said, should consider providing service on holidays. Seminars on hospitality services could be offered. The Trade Show in '08 should focus on the mining sector.

In response to a question, Kathmann said it went without saying that he fully supported the development of a five-year strategic plan with the participation of all members.

At the close of the meeting, a motion from the floor read: "That the Chamber publicly recognize Edgar Lavoie for the work done in promoting mining in the area." The motion referred to a series of articles in the Times Star this past season. The motion was adopted unanimously.

The meeting adjourned about 9:25.

Stephane Parent thanks Larry Joy, Executive Assistant to Thunder Bay-Superior North MPP, Michael Gravelle

Chapter 6 : Year 2007

Muskeg Tours 2009

Attached to the Ministry of Northern Development & Mines

Activities of Ontario Geological Survey

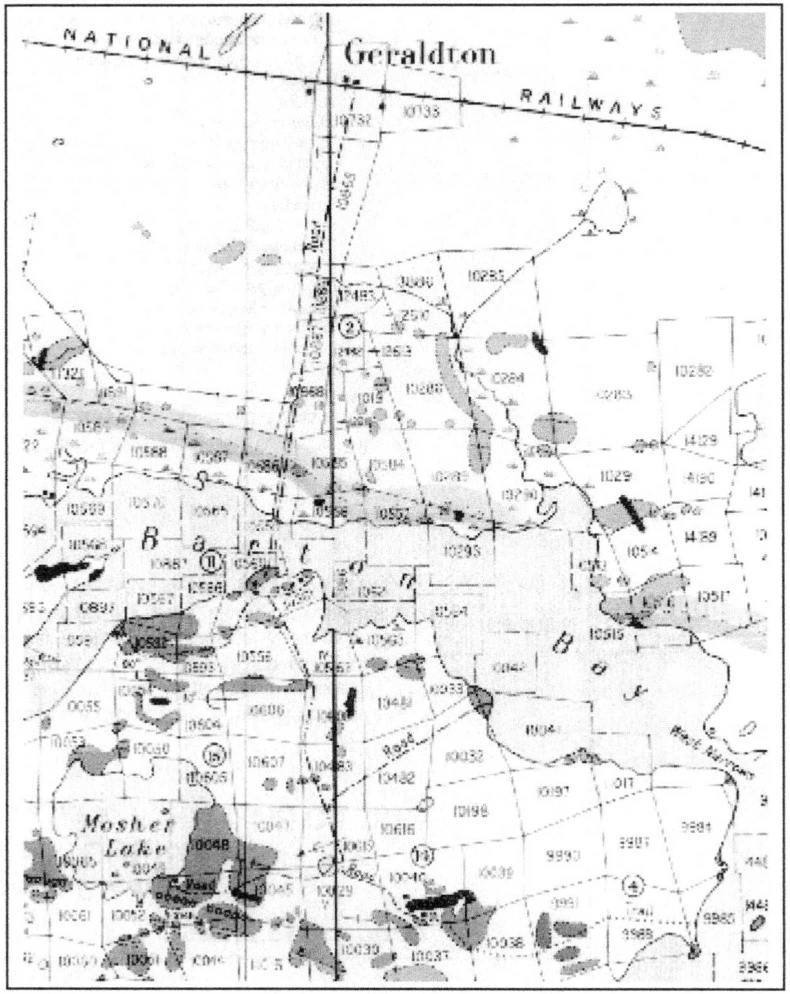

Section of Map 44d, "Little Long Lac Gold Area", ODM 1935

SELECTIONS FROM REPORT OF ACTIVITIES, 2007

Resident Geologist Program
Thunder Bay North District

by M.C. Smyk, G.D. White, M.A. Puumala, P. Hinz and C.L. Komar

INTRODUCTION

The Thunder Bay North District encompasses Lake Nipigon and extends over 500 km north from Highway 11 to the Hudson Bay Lowland (Figure 1). The following communities are situated within the district: Angling Lake (Wapakeka First Nation (FN)); Armstrong; Beardmore; Bearskin Lake FN; Big Trout Lake (Kitchenuhmaykoosib Inninuwug FN); Cat Lake FN; Fort Hope (Eabametoong FN); Geraldton; Gull Bay FN; Jellicoe; Kasabonika FN; Kingfisher Lake FN; Lansdowne House (Neskantaga FN); Longlac; Macdiarmid; Nakina; North Caribou Lake (Weagamow FN); Osnaburgh House (Mishkeegogamang FN); Pickle Lake; Summer Beaver (Nibinamik FN); Webequie FN; and Wunnummin Lake FN.

. . . .

MINING AND EXPLORATION ACTIVITY

. . . .

Seventy-six exploration programs were undertaken in the Thunder Bay North District in 2007 (Table 3). Gold was the most sought-after commodity, followed by platinum group metals, base metals, uranium and diamonds.

. . . .

Abbreviations

AEM	Airborne electromagnetic survey	IP	Induced polarization survey
AM	Airborne magnetic survey	Lc	Linecutting
ARA	Airborne radiometric survey	Met	Metallurgical testing
Beep	Beep Mat survey	OD	Overburden drilling
Bulk	Bulk sampling	ODH	Overburden drill hole(s)
DD	Diamond drilling	PEM	Pulse electromagnetic survey
DDH	Diamond drill hole(s)	PGM	Platinum group metals
DGP	Down-hole geophysics	Pr	Prospecting
GC	Geochemical survey	RES	Resistivity survey
GEM	Ground electromagnetic survey	Samp	Sampling (other than bulk)
GL	Geological survey	Seismic	Seismic survey
GM	Ground magnetic survey	SP	Self-potential survey
GRA	Ground radiometric survey	Str	Stripping
Grav	Gravity survey	Tr	Trenching
HLEM	Horizontal loop electromagnetic survey	UG	Underground exploration/development
HM	Heavy mineral sampling	VLEM	Vertical loop electromagnetic survey
IM	Industrial mineral testing and marketing	VLFEM	Very low frequency electromagnetic survey

. . . .

ADVANCED EXPLORATION

Roxmark Mines Limited continued a surface program to further study the extent of the molybdenum mineralization along trenches No. 1, 2, 4, 5 and 6 on their Nortoba–Tyson molybdenum property near Beardmore (Roxmark Mines Limited, Management's Discussion and Analysis, November 29, 2007; see document filed with SEDAR® (http://www.sedar.com)). Geological mapping, channel sampling and assaying was carried out. A second bulk sample to extract molybdenum ore from these trenches is planned. Another surface program was started to further explore the gold-bearing veins on the property. The work was directed on veins No. 1, 1a, 2 and 7 and included mechanical stripping, mapping, channel sampling and assaying prior to drilling. Several grab samples taken by Roxmark in 2006 from the bottom of the previous bulk sample area returned values ranging from 0.17 to 7.64 ounce Au per ton.

. . . .

Major Exploration Projects

Kodiak Exploration Limited acquired and began exploration on their Hercules gold property, north of Jellicoe, in 2006. Kodiak discovered and verified high-grade gold mineralization in a series of quartz-carbonate vein-bearing shear zones that cut through granodiorite and metavolcanic rocks. Surface stripping in 2006 and 2007 exposed several gold-bearing structures, consisting of quartz vein and stockwork systems, with a combined strike length of more than 5 km. Individual veins are up to 9 m wide and 1000 m long, and are enclosed within a wide envelope of intensely sheared and altered (sericite, chlorite, quartz and hematite) rock (Kodiak Exploration Limited, Management Discussion and Analysis, November 29, 2007; see document filed with SEDAR® (http://www.sedar.com)). In August 2007, Kodiak prospectors discovered the Golden Mile, a major goldmineralized structure 2 km west of, and parallel to, the WL-Penelton-Yellow Brick trend. Surface trenching exposed the Golden Mile over more than 2 km, with a geophysically inferred strike length of more than 4 km. Continuous gold mineralization has been traced for more than 365 m along the Golden Mile structure; visible gold is frequent and conspicuous within this interval. Of the grab samples taken along this zone to date, 41% exceeded 10 g/t Au, and channel samples taken every 20 m along this zone averaged 20.2 g/t Au over an average width of 3.8 m, including one that assayed 32 g/t Au over a width of 11.6 m. Two drill holes beneath this sample also returned good gold values, including 6.89 g/t Au over 4.0 m in hole HR07-09. Financed by a bought-deal equity offering that raised $55 million, Kodiak started a 60 000 m diamond-drilling program to test the Golden Mile, Yellow Brick Road and other gold-mineralized structures. Kodiak has also embarked upon an aggressive strategy to control as much of the Beardmore-Geraldton belt as possible. As a result of staking, joint-ventures and options with private landowners, Kodiak is now the dominant landholder in the area, with a property that covers approximately 1400 km^2 of favourable stratigraphy and structure in an area measuring approximately 150 km long and up to 40 km wide (ibid).

. . . .

Sage Gold Inc. completed 13 diamond-drill holes, totaling approximately 4000 m, on it Jacobus copper-nickel property, north of Jellicoe (Sage Gold Inc., Management Discussion and Analysis, February 21, 2008; see document filed with SEDAR® (SEDAR Home Page)). Current drilling was designed to test lateral and down-dip extensions of a historical resource, as well as to explore the continuity of gold mineralization discovered during the summer 2007 drilling campaign. Sage also continues to explore the possible strike extension of Kodiak Exploration Limited's Golden Mile gold-bearing vein system on the eastern portion of the Jacobus claim block. Sage completed the interpretation of an aeromagnetic survey that was completed in December 2007. Line-cutting, surface stripping, channel sampling and drilling are planned to explore the gold potential of the eastern part of the property.

Sage also completed an induced polarization (IP) survey on the nearby Onaman base metal property, in addition to an airborne electromagnetic (EM) and magnetometer survey. The geophysical surveys were designed to further extend the copper-silver-gold mineralization associated with the Lynx 1, 2 and 3 zones. In addition, Sage discovered a new zone of volcanogenic massive sulphide (VMS) mineralization west of the Lynx zones. This new zone (Lynx #4) was tested by IP and will be drilled in 2008. Sage has been trenching and sampling a 1.5 km long geophysical conductor (The Big M), which occurs on the western side of the Onaman property. Sage intends to drill-test this conductor to approximately 500 m in vertical depth (ibid).

Roxmark Mines Limited drilled their Northern Empire Mine gold property in Beardmore. During the summer of 2007, a 26-hole program, totaling 2711 m, was designed to delineate the extent of the Contact and Power zones above the diabase sill, in the eastern extension of the main vein, 700 to 912 m east of the shaft (Roxmark Mines Limited, Management's Discussion and Analysis, November 29, 2007; see document filed with SEDAR® (http://www.sedar.com)). All holes intersected the Power and Contact zones, with gold mineralization persisting over a strike length of 212.5 m. A surface exploration program was also initiated on the recently acquired Amedé properties, west of Beardmore, particularly on the Summers and Princess lakes properties. This included data compilation, trenching, channel sampling and geological mapping, prior to diamond drilling (ibid).

East West Resource Corporation and **Eyeconomy Holdings PLC** continued to work on their Marshall Lake base metal property, west of Nakina (East West Resource Corporation, Management Discussion and Analysis, December 28, 2007; see document filed with SEDAR® (SEDAR Home Page)). The Marshall Lake zone has a historical, drillindicated resource of 2 211 000 tons grading 1.22% Cu, 4.2% Zn, 2.45 ounces Ag per ton and 0.012 ounce Au per ton (*Canadian and American Mines Handbook*, 2006–2007, p.305). In 2007, exploration continued on the Teck Hill, North Gazooma, Cherry Hill (Teck Cominco Limited leases) and a new area east of Teck Hill (N.W.T. Copper Mines Ltd. leases). Two copper-zinc occurrences were trenched with a backhoe on the North Diabase and North Copper showings on the N.W.T. Copper Mines claims. Geophysical surveys continued on the Main Billiton grid and areas east of Teck Hill and North Gazooma. Five drill holes were completed on the Gazooma area in September: 1 on Cherry Hill; 1 on the Teck Hill claim; and 3 on Teck Cominco leases, for a total of 1447 m in 10 holes. The company increased the size of the Marshall property by adding 161 claim units to the western side of the Marshall Lake property following the encouragement received from the airborne EM survey, and prospecting. This enlarges the property to 965 claim units, or 60 square miles. A large drilling program was planned to start in January 2008 (East West Resource Corporation, Management Discussion and Analysis, December 28, 2007; see document filed with SEDAR® (SEDAR Home Page)).

. . . .

Premier Gold Mines Limited signed a Letter of Intent with **Roxmark Mines Limited** to acquire an interest in a portion of Roxmark's extensive land holdings in the Geraldton area (Premier Gold Mines Limited, News Release, September 24, 2007) and began with a drilling campaign on the past-producing Little Long Lac Mine gold property. This was followed by a drilling campaign at the past-producing Magnet Mine gold property where Premier announced that 2 distinct, gold-mineralized zones were intersected, confirming the presence of significant veining and gold mineralization in the target horizon, including visible gold (ibid, January 24, 2008). The current drilling is part of a major $2.5 million exploration program within a project area that covers approximately 10 km.

Premier also signed a letter of intent to sell its interest in the Pickle Crow Gold Project at Pickle Lake to an unrelated privately held Ontario company, **PC Gold Inc.**, which has been created specifically to consolidate ownership interests in the project (ibid, November 05, 2007). PC Gold intends to raise financing and implement a major program of exploration and development on the project, with the intent of returning the past-producing Pickle Crow Mine to production, and to eventually complete a transaction to become a public issuer. The project also hosts a new mill built by a private partner in the project, who will also be consolidating its interest into PC Gold.

. . . .

PROPERTY EXAMINATIONS

. . . .

Lafontaine–Rentz Property

The Lafontaine–Rentz property is located east of Lake Nipigon in the northwestern portion of Sandra Township, approximately 12 km north-northwest of Beardmore. The property consists of 6 claims (58 units) along a westerly trend and is bounded by Kodiak Exploration Limited's property to the east, north and south, and properties held by Laurion Minerals Exploration Inc. / Geoinformatics Exploration Canada Ltd. and Roxmark Mines Limited, respectively, to the west. Access to the Lafontaine property is via the 72 Road North from Highway 580 (Poplar Lodge Road). An all-weather gravel road, upgraded by Roxmark Mines Limited in late 2006 to access its Nortoba–Tyson gold-molybdenum property for bulk sampling, traverses the central portion of the Lafontaine claim group.

The Lafontaine–Rentz property lies within the southeastern part of the Wabigoon Subprovince along the Paint Lake fault / deformation zone (PLDZ). This major east-trending fault structure separates the older Onaman–Tashota metavolcanic terrane to the north from the main Beardmore–Geraldton metavolcanic-metasedimentary belt to the south. The fault, delineated by topographic lineaments in Sandra and Dorothea townships (Mackasey 1975), is expressed as an intense zone of deformation that appears to be bounded by 2 distinct, subparallel linear faults (Hart et al. 2002). The western extension of the PLDZ is well-exposed along the eastern shore of Lake Nipigon, north of what is known as the Devil's Walk, in north-central Dorothea Township. Rocks are highly strained, as exhibited by the absence of primary volcanic features such as pillows, and by the development of a strong mylonitic fabric parallel to the PLDZ (Hart et al. 2002; Mason et al. 1999). An examination of the 1988 Tashota–Geraldton Aerodat Airborne Survey data shows magnetic contours that trace the PLDZ across the north-central portions of Sandra and Dorothea townships, westward to Devil's Walk. This is displayed as a linear series of magnetic "highs" and depressions (OGS 1989).

The exploration history of the Lafontaine–Rentz property is limited, since most of the claim group is underlain by Mesoproterozoic Nipigon diabase sills. An examination of the area by staff of the Thunder Bay North Resident Geologist's office confirmed the extent of the diabase in this area. This brown-weathering rock typically consists of dark grey, equigranular, medium- to fine-grained gabbro with a well-developed diabasic texture. In 1988, a diamond-drilling program was conducted by Edda Resources Inc. (assessment files, Thunder Bay North District, Thunder Bay). It consisted of 7 drill holes across the eastern portion of the Lafontaine claim group, in the area of Little Dawson Lake. Results from this drilling program indicated that the diabase sill varies in thickness from approximately 100 m to less than 200 m. In all cases, drill core recovered from below the diabase consists of Archean mafic to intermediate metavolcanic flows and tuffs, felsic tuffs and porphyry, and metasedimentary rocks. Drill sections of sheared and/or brecciated, chlorite-sericite-altered, intermediate to felsic metavolcanic rocks, containing 1 to 2% fine pyrite, were noted in several of the holes. Assay results from 4 of the 7 drill holes, of samples collected within the altered metavolcanic rocks, varied from 0.75 g/t Au over 1.46 m (hole BRD88-04) to 1.37 g/t Au over 1.43 m (hole BRD88-7). In Irwin Township, 10 km east of the Lafontaine property along the PLDZ, recent diamond drilling by Ontex Resources Ltd. on its Brookbank deposit has outlined an indicated and inferred resource estimate of 2.6 Mt grading 7.4 g/t Au (*Canadian and American Mines Handbook*, 2007–2008, p.435). This demonstrates the economic significance of this structure.

The exploration potential for gold mineralization along the PLDZ in the area of the Lafontaine–Rentz property is high. Although the diabase cover has restricted exploration in the past, recently developed, deep-penetrating airborne geophysical systems can "see through" this cover. Well-defined targets can then be followed up with drilling.

Roxmark–Lafontaine Property

The Roxmark–Lafontaine property is located in the southwestern corner of Summers Township, immediately west of Beardmore. Access is via the waste disposal road north of the Blackwater River bridge, which tracks southwest for approximately 4 km to the central portion of the property. Roxmark Mines Limited optioned the 59-claim (97 units) group from the Lafontaine partners of Beardmore late in 2007 and initiated a stripping, trenching, sampling and geological mapping program (News Release, Roxmark Mines Limited, November 08, 2007). The property lies along the western strike extension of the company's Northern Empire Mine, in Beardmore. The property was previously described by Smyk et al. (2006) and was revisited in 2007 by staff of the Thunder Bay North District in order to document recent exploration activity carried out by the vendors and optionors.

The southwestern portion of Summers Township was the focus of extensive exploration activity during the 1930s. From 1935 to 1939, a considerable amount of surface work, including the sinking of a 24.4 m (80-foot) shaft and over 3048 m (10 000 feet) of diamond drilling, were completed on the Buffalo–Beardmore Gold Mines Ltd. occurrence in the south-central portion of the current Roxmark property (Resident Geologist files, Thunder Bay North District, Thunder Bay). Historical descriptions report gold values up to 0.42 ounce Au per ton across 2.64 m (8.69 feet) associated with quartz veins in sulphide (pyrite and arsenopyrite)-chert-magnetite banded iron formation. Numerous other vein-related gold occurrences were discovered and worked until the early 1940s. Hart et al. (2002) documented and described many of these showings, including the Anglo–Beardmore (located 7 km northeast of the Buffalo–Beardmore occurrence) and the C.J. Morgan occurrences, located along the Princess Chain of Lakes in the northwestern portion of the property. Very little exploration activity occurred in the area until 1981 when Gold Field Resources Canada Limited conducted extensive ground magnetometer and very low frequency electromagnetic (VLF-EM) surveys over a large grid and completed 4 diamond-drill holes, totaling 484.7 m, in the northern part of the property. In 1984, Thorco Gold Finders Ltd. completed mapping, stripping, ground geophysical surveys and diamond drilling over the Anglo–Beardmore occurrence. Following an airborne magnetometerelectromagnetic survey and a reverse-circulation drilling program conducted by Legion Resources Ltd. in 1986 in southwestern Summers Township (Hart et al. 2002), the property was subsequently optioned by Gold Dragon Resources Ltd. and Glen Auden Resources Inc. in 1989. This property became part of a larger package of ground extending westward across 2 townships into Lake Nipigon. These companies conducted reconnaissance-scale mapping and sampling, mechanical stripping and a VLF-EM survey over claims covering the Gold Fields and Morgan occurrences in the Princess Chain of Lakes area. From 1990 to 1994, the Lafontaine partners completed extensive groundwork, involving prospecting, stripping and sampling, in the southern portion of the claim group. In 1996 the property was optioned to Explorations Minière du Nord Ltée. The company carried out extensive stripping and sampling, and completed 24 diamond-drill holes, totaling 3082 m. Much of the exploration work was concentrated along the Empire Fault in the central portion of the property and to the south in the vicinity of the Buffalo–Beardmore and Anglo–Beardmore occurrences (Hart et al. 2002). Extensive units of folded, sulphidefacies banded iron formation were targeted. In 2007, the 59-claim property (divided by the Lafontaine partners into 3 separate properties) was optioned by Roxmark Mines Limited (News Release, Roxmark Mines Limited, November 08, 2007). The company initiated stripping, sampling and mapping of the old trenches and occurrences along the southern half of the property in late 2007.

The Roxmark–Lafontaine property lies almost entirely within the thickest portion of the western part of the Southern Metavolcanic Sub-belt (SVB), which is part of the main Beardmore–Geraldton belt in the southeastern Wabigoon Subprovince. It is situated immediately southwest and along strike of the Northern Empire Mine, which produced 149 493 ounces of gold at a grade of 0.35 ounce Au per ton from 1934 to 1941 (Mason and White 1986). The mine is characterized by a series of gold-bearing, *en echelon* quartz veins containing variable amounts of pyrite, arsenopyrite, pyrrhotite and chalcopyrite, hosted by carbonatized, sheared metavolcanic rocks and interflow iron formation of the SVB. Economic mineralization within this host rock has occurred adjacent to the Empire Fault and at the northern contact of the SVB with the Southern Metasedimentary Belt (SSB). Quartz veins are localized by brittle fractures in the deformed iron formation (Hart et al. 2002). The Empire Fault can be traced across the central portion of the Roxmark–Lafontaine property. Many of the known gold occurrences along this structure and to the south are similar to those at the Northern Empire Mine.

A structural interpretation of this area of Summers Township was first discussed by Benedict and Titcomb (1947), who recognized the Empire Fault in the underground workings of the Northern Empire Mine. The fault is marked by a thin fault gouge of black graphite bordered by a few feet of sheared metavolcanic rock (Benedict and Titcomb 1947). A reinterpretation of the area by Hart et al. (2002) during the Phoenix Bedrock Mapping Project resulted in the separation of the original Empire Fault into 2 distinct faults, the Empire Fault and the Princess Lake Fault. In eastern Summers Township, the Empire Fault trends approximately 060° and extends along the metavolcanic metasedimentary contact adjacent to the Northern Empire Mine. Further west into the Roxmark–Lafontaine property, the fault bends to the south and passes obliquely through the SVB, trending toward Lion Bay in Lake Nipigon. The Princess Lake Fault extends parallel to the Princess Chain of Lakes at 060°. This fault follows the southern edge of the SSB and roughly traces the northern boundary of the Roxmark claim group. An increase in strain, including intense local shearing within the metavolcanic-metasedimentary host rocks, defines the fault in this area (Hart et al. 2002).

The southern limit of the Roxmark claim group is marked by the Blackwater River Fault, which is an expression of the boundary between the Wabigoon Subprovince and the metasedimentary gneisses of the Quetico Subprovince to the south. Carter (1987), Mackasey (1970), Shanks (1993) and, most recently, Hart et al. (2002) all recognized and mapped this structure. The fault is exposed in a few locations southwest of Beardmore along Highway 11. The rocks generally display an increase in strain, such as intense to extreme flattening of pillows (up to 1:20 elongation ratio) and the development of a strong to intense cleavage within the metasedimentary rocks (Hart et al. 2002).

The main iron formation showing (UTM 428103 E / 5493037 N), visited by staff of the Thunder Bay North Resident Geologist office in 2007, is located near the center of the Roxmark claim group (TB 1194269 and 1194270) and is part of the western extension of the Anglo–Beardmore occurrence (Hart et al. 2002). The chertmagnetite banded iron formation can be followed along strike (240° to 245°) for 440 m, dips 85° to the north and varies from 1 to 12 m in width. Locally, weathering has resulted in extensive gossan development, particularly where sulphidization of the iron formation has occurred. These sections are commonly crosscut by quartz and quartcarbonate veins. The sulphide content at these locations averages 10 to 15%, but can be as high as 40%, and consists of pyrite, pyrrhotite, chalcopyrite and arsenopyrite. Arsenopyrite and pyrite are the dominant sulphides associated with the quartz-carbonate veining. The host rock to the iron formation consists of highly chloritized and carbonatized, massive to pillowed, mafic metavolcanic flows containing 1 to 3% pervasive, fine pyrite. A grab sample of this altered, sulphide-mineralized, massive flow unit collected by Hart et al. (2002) returned 0.08 ounce Au per ton and 158 ppm Cu. In some areas, both the mafic metavolcanic rock and the banded iron formation are tightly folded, sheared and brecciated near the most heavily mineralized sections. Assay results from a diamonddrill hole completed by Explorations Minière du Nord Ltée in 1996 across the central portion of the main banded iron formation returned 3.16 g/t Au over 1.16 m in hole LA-96-12 (assessment files, Thunder Bay North District, Thunder Bay). Selected grab samples of banded iron formation and altered metavolcanic host rocks, collected from across the central portion of the property by Explorations Minière du Nord Ltée in 1996 and 1997, returned values ranging from 2 to 89 g/t Au. Two grab samples collected by Thunder Bay North Resident Geologist Program staff from the main zone of sulphidized, chert-magnetite banded iron formation, containing 20% arsenopyrite, returned assay values of 0.03 and 0.08 ounces Au per ton (Resident Geologist files, Thunder Bay North District, Thunder Bay).

Work initiated by Roxmark Mines Limited in the fall of 2007 concentrated on assessing and sampling the known banded iron formation-hosted gold occurrences and exploring the Empire Fault along its strike extension, west from the Northern Empire Mine. The intersection of this fault structure and the main banded iron formation, both on surface and at depth, is a primary target for gold mineralization.

RECOMMENDATIONS FOR EXPLORATION

Gold in Felsic Intrusions

Although felsic intrusions (e.g., tonalite, trondhjemite, granodiorite, granite, quartz diorite, etc.) in and around greenstone belts have never garnered as much gold exploration as the belts themselves, the recent discovery of the Golden Mile zone by Kodiak Exploration Limited within the Elmhirst granodiorite pluton on their Hercules property, north of Jellicoe, should prompt a re-evaluation of felsic intrusions as potential hosts to gold mineralization. As shown by Colvine et al. (1988), for example, large felsic intrusions, while accounting for over one-quarter of the Abitibi greenstone belt lithologies, are not represented at all (in statistically significant amounts) as hosts to gold mineralization at either a mine- or deposit-scale. Most Archean lode gold deposits are hosted by supracrustal rocks and minor, post-volcanic intrusions within the supracrustal assemblages. However, notable exceptions do exist, including deposits in the Bourlamaque Batholith, Quebec (e.g., Ferdeber deposit; Vu 1990); the Wawa area (e.g., Renabie; Callan and Spooner 1998; Eagle River); and in Western Australia (e.g., Woodcutters-Lady Bountiful; Mason 2004). Gold is typically hosted by quartz-carbonate veins, which are localized by brittleductile shear zones that crosscut the intrusion and have a protracted history of movement and dilation. Veins and/or permeable shear zones are usually enveloped by altered wall rocks. Alteration assemblages may consist of mica (e.g., sericite, muscovite, biotite), chlorite, sulphides (e.g., pyrite), oxides (e.g., hematite) and epidote, and are dictated by fluid and wall rock chemistry, temperature and metamorphic conditions.

The Elmhirst Lake, Coyle Lake and Kaby Lake stocks range in composition from tonalite to quartz diorite and granodiorite (Mackasey and Wallace 1978; Stott et al. 2002). U-Pb dates from the Elmhirst Lake and Kaby Lake stocks are 2736.1 ± 1.5 Ma and 2734 ± 1 Ma, respectively (Stott et al. 2002). Although they intrude metavolcanic rocks of the Elmhirst-Rickaby assemblage (ca. 2740 Ma), they are just slightly younger and may, in part, be synvolcanic with this assemblage. They have experienced regional metamorphism as well. Mineral occurrences in Elmhirst and Rickaby townships have been documented by Mackasey and Wallace (1978) and Parker (1996). Significant felsic intrusion-hosted, shear-zone-related gold occurrences are shown on Figure 5.

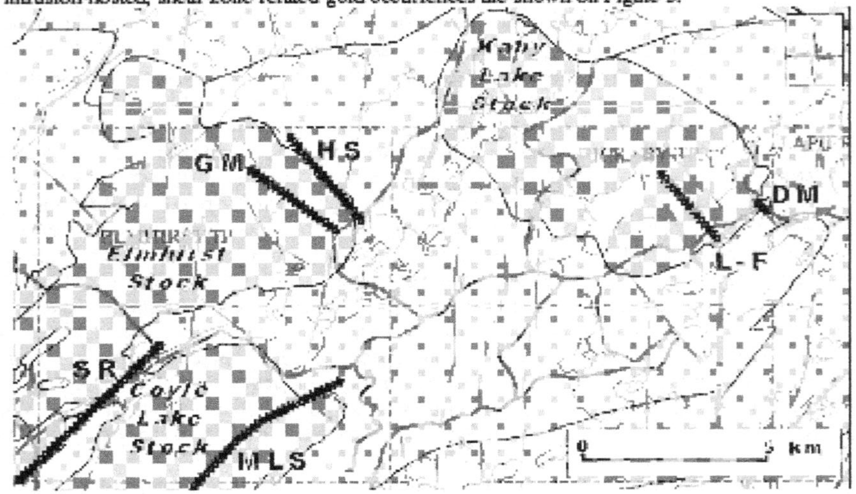

Figure 5. Geology and gold occurrences in Elmhirst and Rickaby townships. Gold-bearing structures (thick black hachured lines) in granodiorite (dark grey) stocks: DM – Dikdik (Orphan) Mine; GM – Golden Mile; HS – Hercules Shear; L-F – Larson-Foisey; MLS – Mud Lake Shear; SR – Sturgeon River. Solid, straight line segments are faults and other structures, which may host diabase dykes. Country rocks are metavolcanic rocks of the Onaman–Tashota greenstone belt. Geology *after* Stott et al. 2002.

The Hercules property is located in Elmhirst Township, approximately 30 km east of Lake Nipigon. Channel sample assay results (News Release, Kodiak Exploration Limited, October 01, 2007) from the Golden Mile occurrence (11.6 m grading 32.96 g/t Au), initiated a major staking rush in the area. Kodiak, along with several other junior exploration companies, staked much of the southern and central portions of the Onaman–Tashota and Beardmore–Geraldton belts, respectively. Subsequent news releases from ongoing work by Kodiak Exploration Limited, which has included extensive stripping, channel sampling, mapping, airborne and ground geophysical surveys, and drilling, have defined a quartz-vein-bearing zone along a northwest-trending fracture system within the Elmhirst Lake pluton. The Golden Mile vein exhibits boudinage, ranges in width from 1 to 12 m and is exposed on surface for over 2 km along a strike of 310°. Channel samples collected by the company near the central portion of the zone confirmed the presence of a continuous, gold-mineralized zone averaging 20.20 g/t Au over a strike length of 400 m with an average width of 3.8 m (ibid, November 29, 2007). Additional work conducted by Kodiak Exploration Limited indicated the potential existence of other, parallel, gold-bearing vein systems within the intrusion to the north of the Golden Mile structure.

The significance of this new discovery is realized in the fact that very little exploration had been undertaken until recently within the Elmhirst Lake and nearby plutons. Much of the past exploration has been limited to the contact zone between the granodiorite and the intermediate to felsic metavolcanic rocks. The discovery of other northwesttrending fracture systems within these 3 plutons provides new, previously unrecognized gold targets. Kodiak Exploration Limited has had some success in delineating these fracture systems, by employing very detailed and sensitive magnetometer surveys, using the Golden Mile signature as an example. A linear magnetic depression along a coincident magnetic "high" defines the Golden Mile zone along the known exposure and for a further 2.5 km to the northwest (S. Roach, Kodiak Exploration Limited, personal communication, 2007).

The Foisey and Larson gold occurrences are also hosted in northwest-trending fracture systems (315 to 320°) within the Kaby Lake stock (Mason et al. 1989; Mason and White 1992; Parker 1996). The main Foisey occurrence, which is located between Kaby and Dilla lakes, can be traced for 140 m along strike and varies in width from 8 to 10 m. It is characterized by a quartz vein breccia within silicified, medium- to coarse-grained granodiorite containing 1% disseminated, fine-grained pyrite. Grab samples returned up to 0.62 ounce Au per ton (Resident Geologist files, Thunder Bay North District, Thunder Bay). The Larson occurrence, located southwest of Dilla Lake, is a 30 m wide, sheared, brittle fracture system containing 2 to 10% pyrite, within highly silicified, carbonatized granodiorite. Assay results averaged 0.06 ounce Au per ton over 20 m (Resident Geologist files, Thunder Bay North District, Thunder Bay). Both occurrences are characterized by pink alteration selvages next to veins and mineralized fractures; this was originally described as potassic feldspar alteration. The same pink to red alteration was recently observed in the host granodiorite (Elmhirst Lake stock) along the northern portion of the Golden Mile zone on the Hercules property. Detailed mineralogical analysis of the altered granodiorite by Kodiak Exploration Limited has identified this pink "staining" as hematite (S. Roach, Kodiak Exploration Limited, personal communication, 2007). Hematite alteration, silicification and coarser-grained granodiorite within the fracture zone, are common to all 3 of these gold occurrences. Low-grade but significant gold mineralization is hosted by this hematitized granodiorite in both the Kaby Lake pluton-hosted occurrences and in the Golden Mile zone (up to 2 g/t Au) (S. Roach, Kodiak Exploration Limited, personal communication, 2007).

The recognition of these gold-mineralized, northwest-trending fracture systems and associated hematite alteration within these felsic intrusions should prompt investigation of other intrusive bodies of similar age and composition within the Onaman–Tashota belt. Lafrance et al. (2004) demonstrated that in the Beardmore–Geraldton belt, adjacent to the Onaman–Tashota belt, gold deposits and gold-bearing ore shoots are parallel to the intersection between D_3 shear zones and map-scale folds. The northwest-trending shear zones that cut these 3 felsic intrusions may be correlative with the D_3, dextral shear zones on the southern side of the Paint Lake Deformation Zone (G. Stott, OGS, personal communication, 2008). They should be considered prime gold targets.

. . . .

Table 6. Mineral deposits not being mined in the Thunder Bay North District in 2007.

Abbreviations

AF	Assessment Files	MLS	Mining Lands, Sudbury
CAMH	*Canadian and American Mines Handbook*	MR	Mining Recorder
GR	Geological Report	NM	*The Northern Miner*
MDC	Mineral Deposit Circular	OFR	Open File Report
MDIR	Mineral Deposit Inventory record	PC	Personal Communication

Deposit Name/ NTS	Commodity	Tonnage-Grade Estimates and/or Dimensions	Ownership Reference	Reserve Reference	Status
Barton Bay (42E10/NW)	Fe	500 000 000 tons (grade unknown)	Rosmark Mines Limited (www.rosmark.com)	Resident Geologist's Files, Thunder Bay North District	Inactive
Brookbank (42E12/NW)	Au	Indicated and inferred resource estimate of 2.6 Mt @ 7.4 g/t Au	Ontex Resources Ltd. (CAMH 2007–2008, p.435)	Ontex Resources Ltd. (CAMH 2007–2008, p.435)	Drilling in 2007; discussions with potential partners ongoing
Central Ontario Range (42L03/NW)	Fe	200 000 000 tons @ ca. 30% Fe		MDC 11, p.407	Inactive
Fox Township (52H09/NE)	Fe, P	AF 414: 3 400 000 tons @ 33.5% Fe, 0.115% P to 500 foot depth AF 415: 5 000 000 tons @		MDC 11, p.379	Inactive
Jacobus (42E13/SE)	Cu, Ni	1 million tons @ 0.84% Cu+Ni	N.W.T. Copper Mines Limited (CAMH 2007–2008, p.403)	CAMH 2006–2007, p.303	Optioned by Sage Gold Inc.; drilling 2006–2007
Key Lake (Lindsley) (42E13/NE)	Au	Indicated resource potential of 200 000 oz per 100 vertical metres	Rosmark Mines Limited (CAMH 2007–2008, p.495)	CAMH 2004–2005, p.368	Exploration last conducted in 1996–1997; Qualifying Report being prepared
Lynx Canyon (42L04/SE)	Cu, Ag	25 000 t @ 1.0% Cu, 300 000 t @ 2.0% Cu, 1.0 opt Ag	N. Cox, M. Holt (optioned to Sage Gold Inc.)	OFR 5680, p.41	Optioned and explored by Sage Gold Inc. in 2006–07

Lynx Canada (42L04/SE)	Cu, Ag	25 000 t @ 4.0% Cu, 300 000 t @ 2.0% Cu, 1.0 opt Ag	N. Cox, M. Holt (optioned to Sage Gold Inc.)	OFR 5630, p.41	Optioned and explored by Sage Gold Inc. in 2006–07
Macleod–Cockshutt–Hardrock mines (42E10/NW)	Au	Open Pit resource: 1.92 Mt @ 0.079 oz Au per ton; Ramp resource: 1.16 Mt @ 0.127 oz Au per ton	Barrick Gold Corporation (formerly Lac Minerals Properties Inc.) (Resident Geologist's files, Thunder Bay)	Amoco Exploration Company of Canada Limited Ore Reserve estimate, 1993 (Resident Geologist's files, Thunder Bay)	Former mine sites rehabilitated; last active exploration in 1996
Magnet Mine (42E11/NE)	Au	69 200 tons @ 0.34 oz Au per ton measured and indicated	Roxmark Mines Limited (CAMH 2007–2008, p.495)	CAMH 2006–2007, p.276	Inactive since 1988; Joint-venture signed with Premier Gold Mines Limited on Geraldton properties, 2007
Marshall Lake (42L05/NE)	Cu, Zn, Ag, Au	2.211 million tons @ 1.22% Cu, 4.2% Zn, 2.45 oz Ag per ton, 0.012 oz Au per ton	East West Resource Corporation–Eveconomy Holdings PLC, in part optioned from N.W.T. Copper Mines Limited (East West Resource Corporation, news releases, June 27, 2005 and October 25, 2006)	CAMH 2005–2006, p.289	Renewed exploration, including drilling 2006–2007
....					
Nortobe–Tyson (52H05/NE)	Mo	72 000 t @ 1.04% Mo	Roxmark Mines Limited (CAMH 2007–2008, p.495)	www.roxmark.com	Advanced Exploration status; approx. 2000 t of ore excavated and milled in 2006 (Roxmark Mines Limited, news release January 16, 2007)
....					
Northern Empire Mine (42E12/SW)	Au	Total probable resource of 65 969 tons @ 0.47 ounce Au per ton	Roxmark Mines Limited (CAMH 2007–2008, p.495)	CAMH 2007–2008 p.495	Mill upgraded and re-opened, surface ore from other Roxmark properties processed in 2005–2006; drilling 2007
....					

Chapter 7 :
Year 2007

Maps for the world-wide mining industry

Maps courtesy of Intierra Mapping

Muskeg Tours 2009

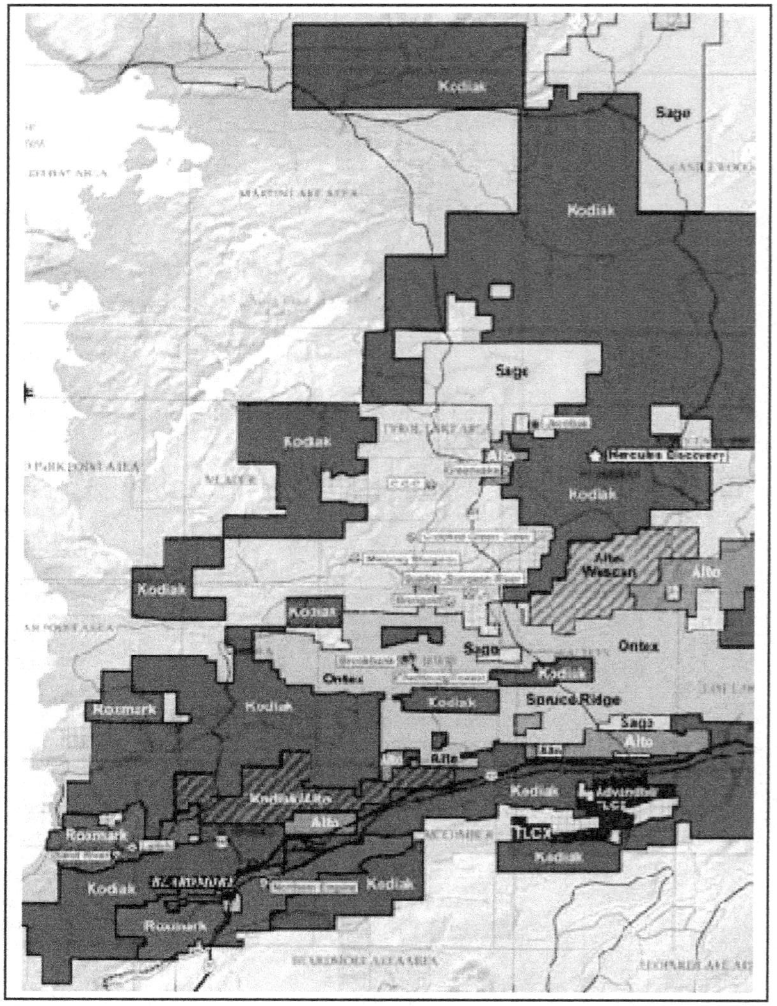

Section of Map "BEARDMORE/GERALDTON : Mining and Exploration Activities", February 2008

Map "BEARDMORE/GERALDTON"

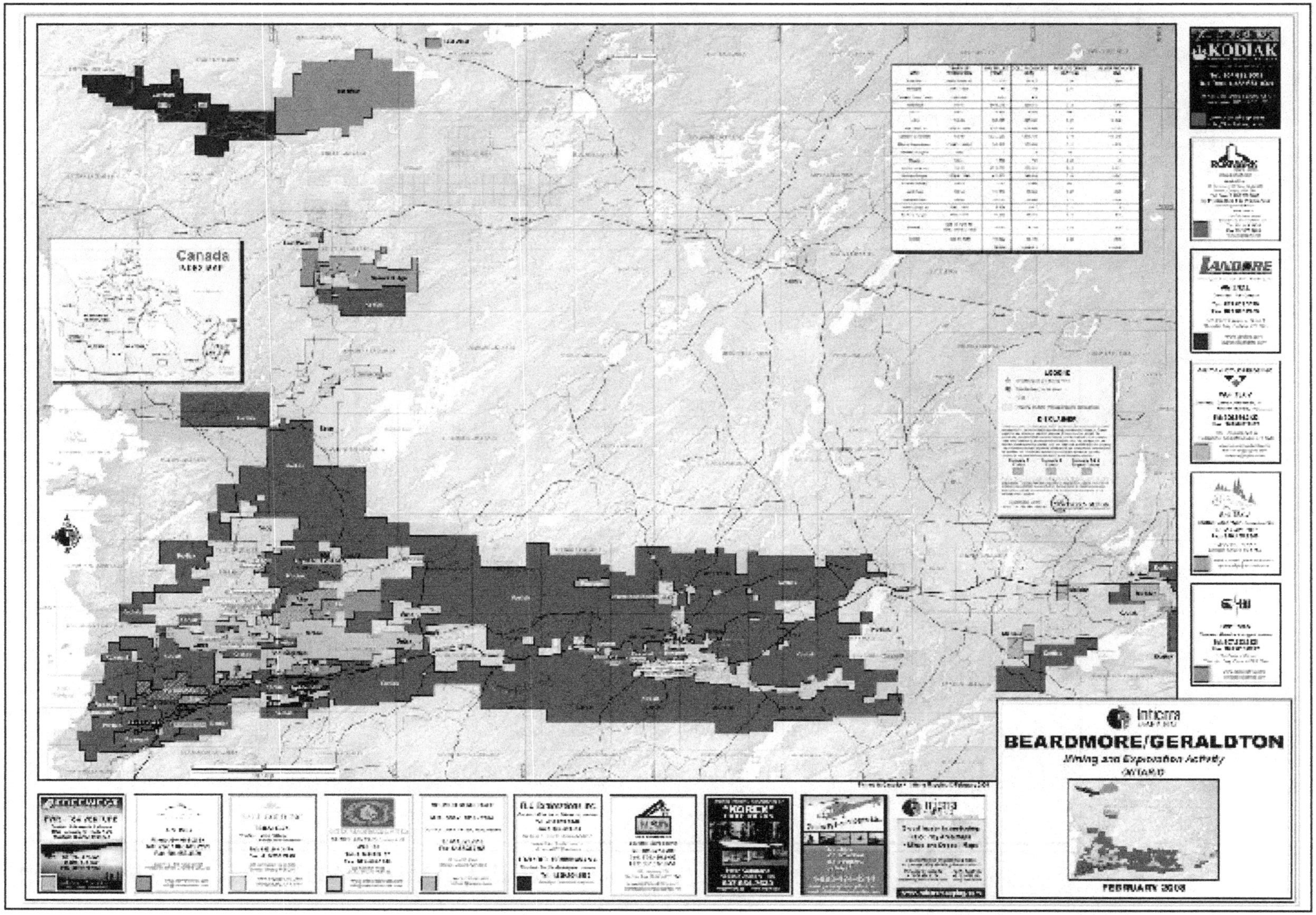

Sections of Map "BEARDMORE/GERALDTON"

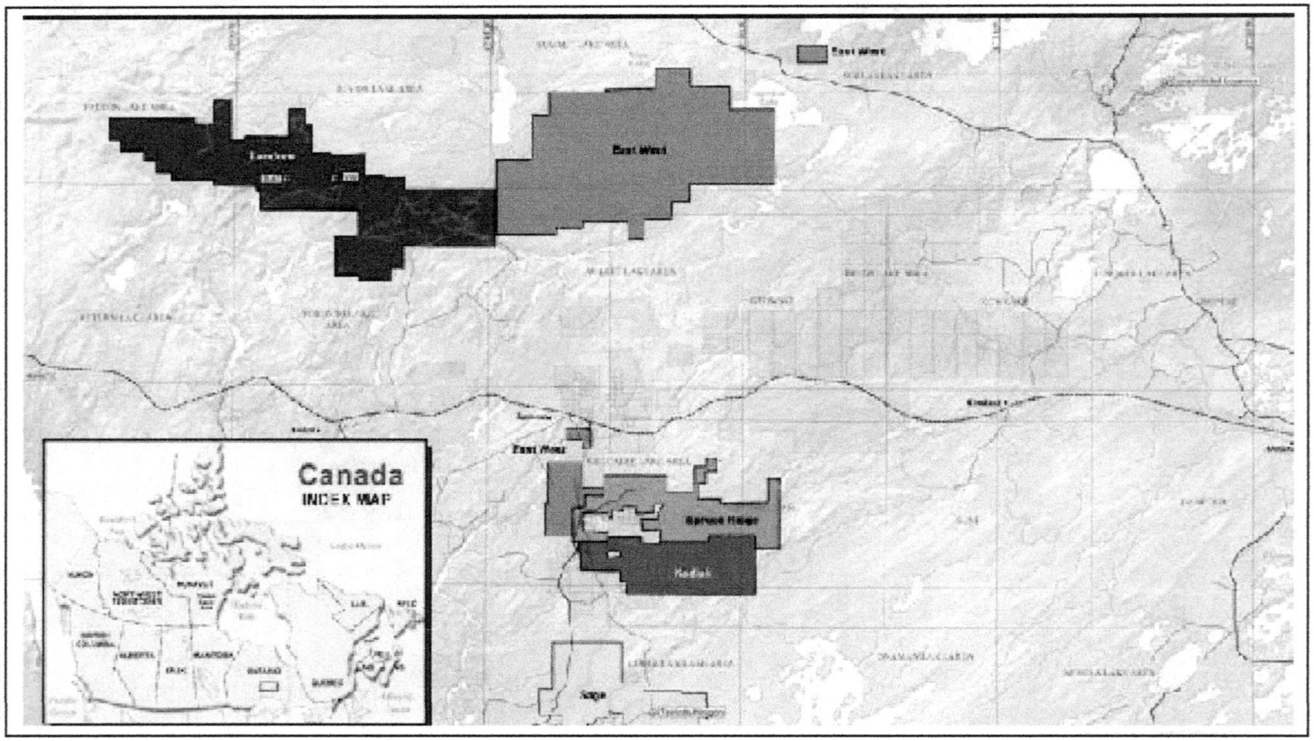

Northeast of Lake Nipigon

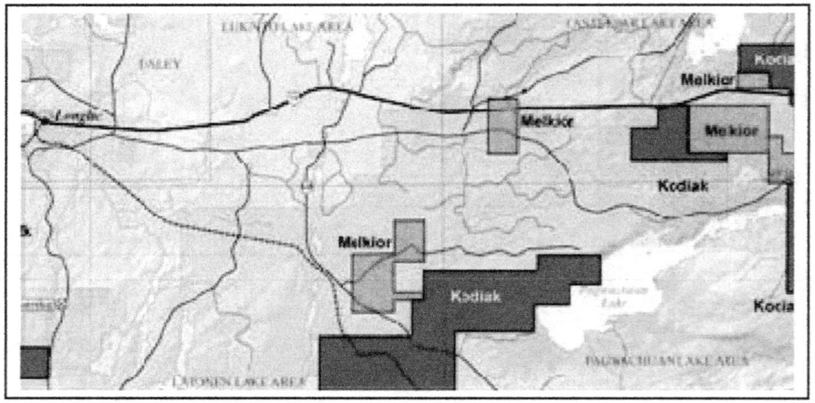

East of Longlac

Map "McFAULD'S LAKE : Mining and Exploration Activity" January 2008

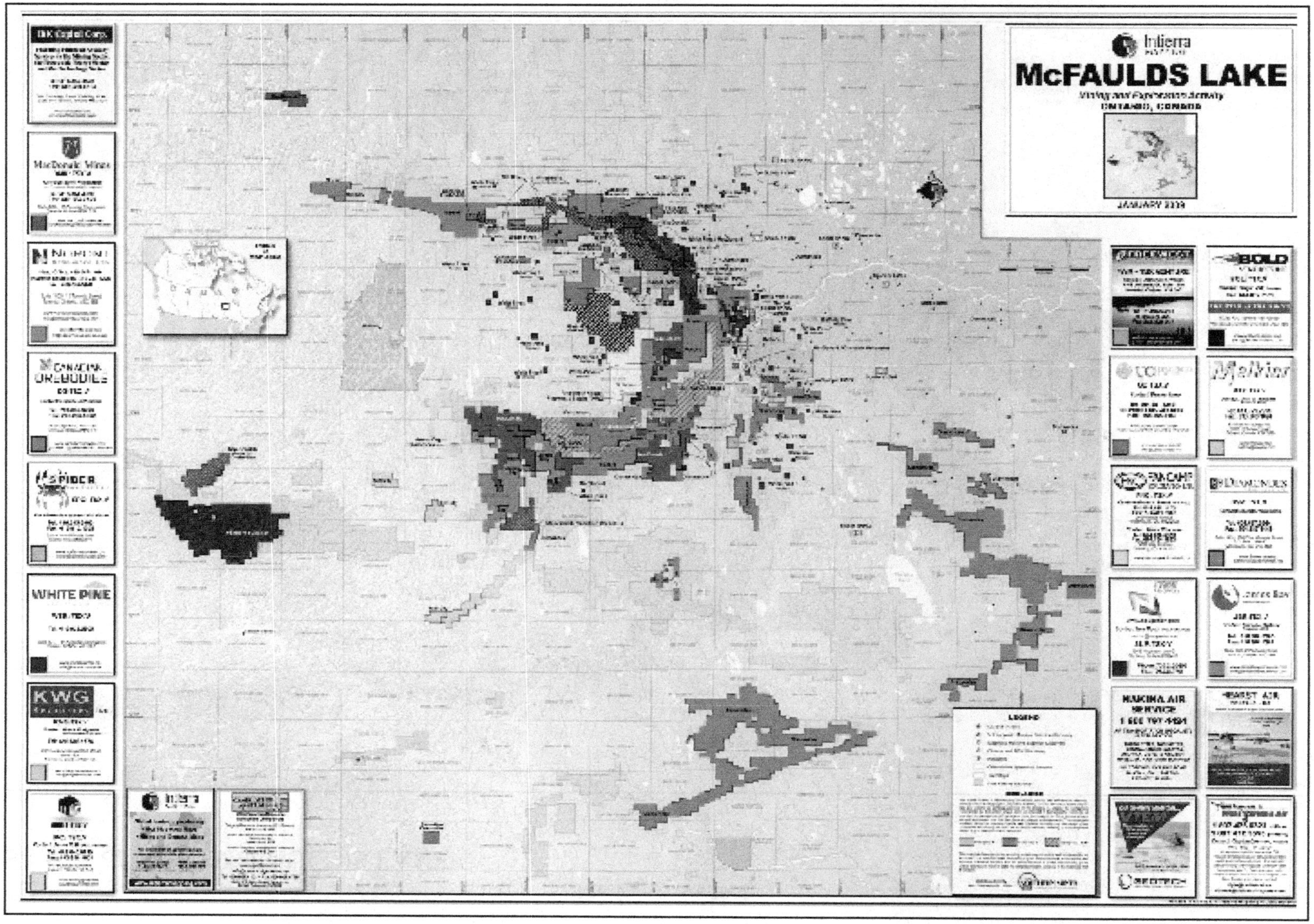

Section of Map "McFAULD'S LAKE"

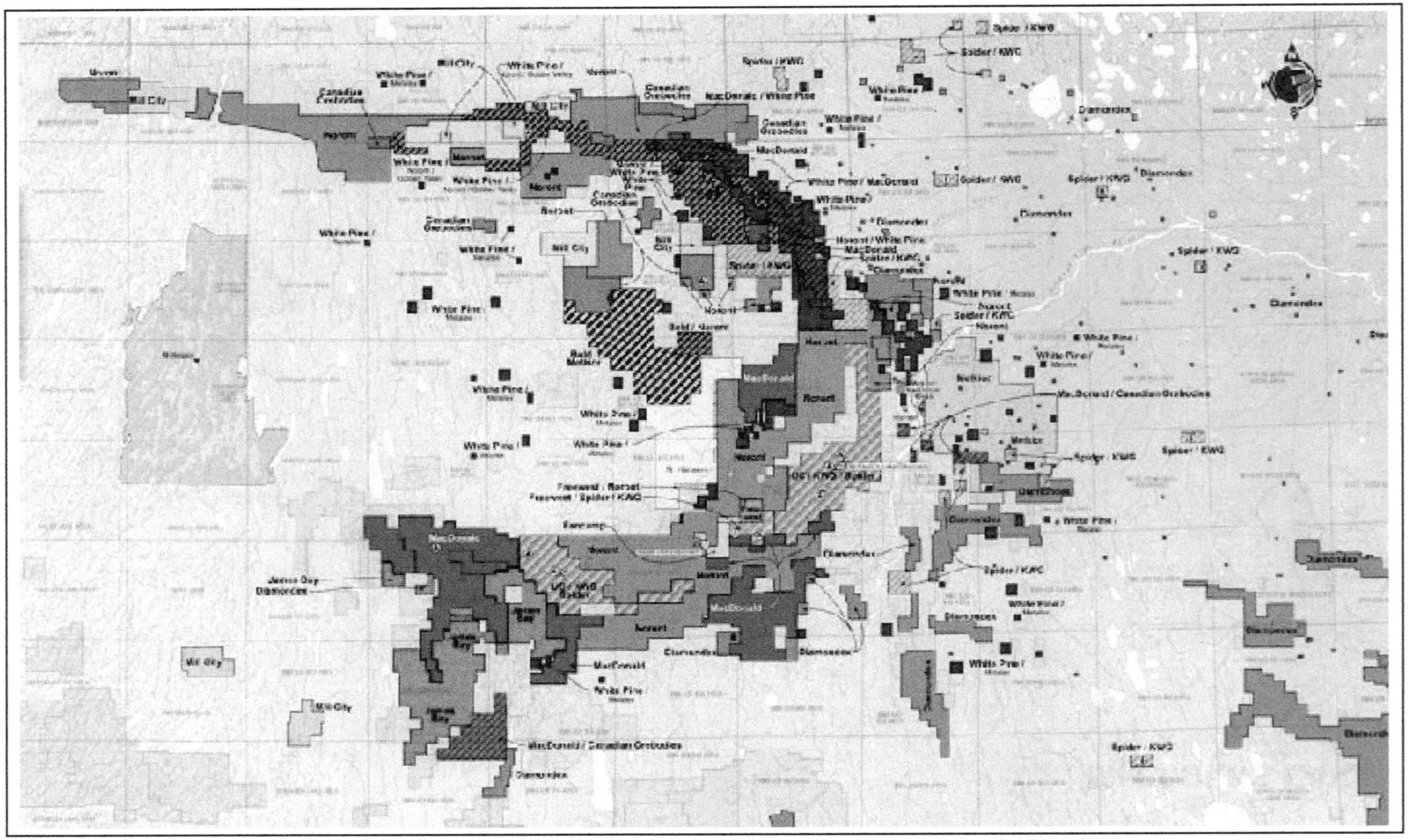

"THE RING OF FIRE"

Section of Map "McFAULD'S LAKE"

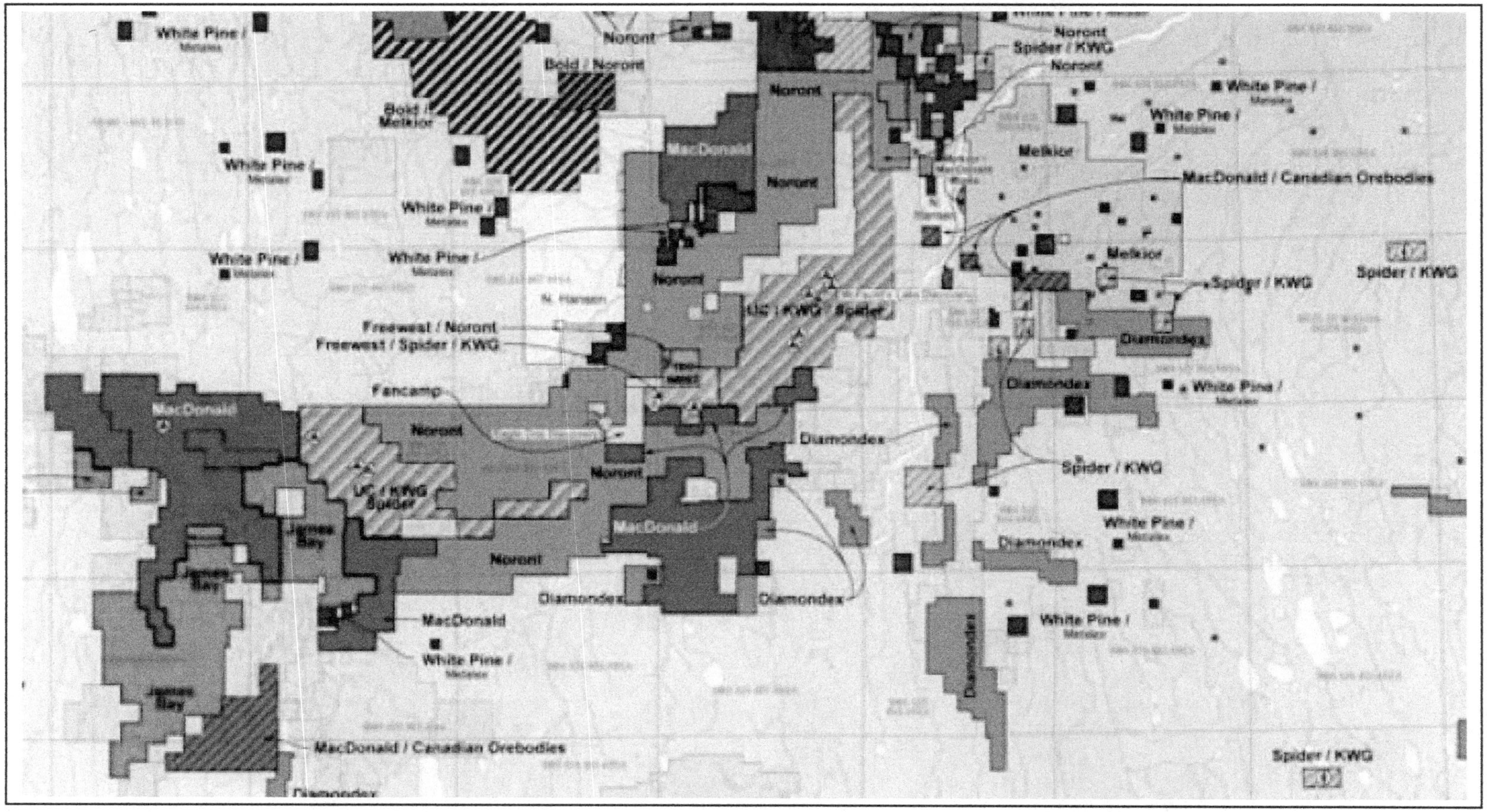

Noront Resources Limited dominates "The Ring"

Year 2008 :

Chapters 1 to 10

Muskeg Tours 2009

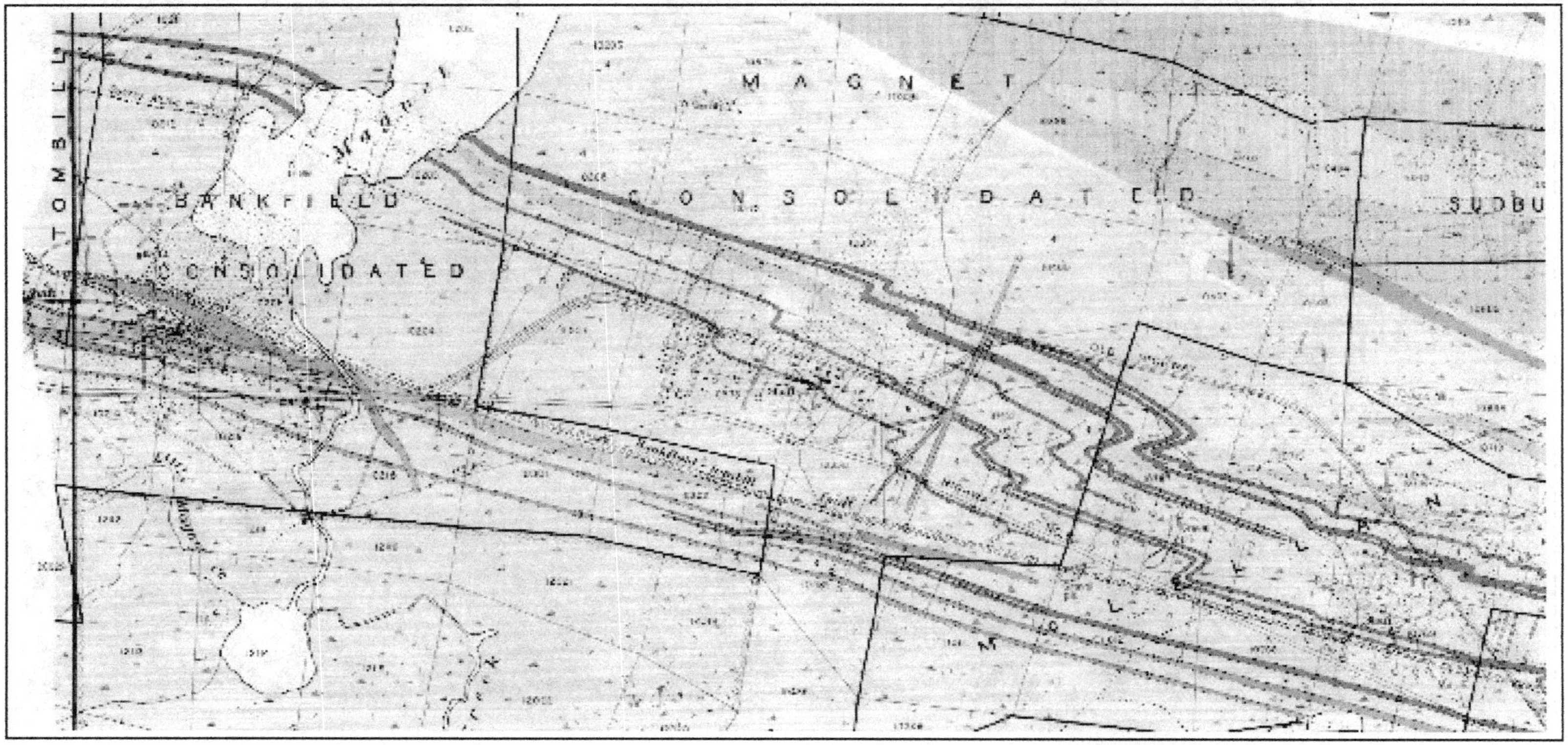

Section of Map 1951-7 ,"Township of Errington", ODM 1951

BEARDMORE-GERALDTON GOLD BELT : Eastern Section (Google Earth View From 40 Miles Altitude)

Beardmore-Geraldton Gold Belt

Even after the Canadian Northern Ontario Railway finished its line between Longlac and Port Arthur (now Thunder Bay North) in 1914, no one, with one exception, appears to have considered the Little Long Lac area as gold country. Little Long Lac, of course, is the translation of the aboriginal name for the lake at Geraldton, Kenogamisis.

The CNOR became the CNR in 1919. In 1931 a prospector, William Smith (later referred to as "Hardrock" Bill Smith), made a strike on a point of Kenogamisis, just south of the narrows that led west into Barton Bay. In the fall, another prospector, Tom Johnson, accompanied by Bob Wells, cruised the lake. They paddled west up Magnet Creek, which emptied into Barton Bay, and on Magnet Lake they found colour. In early May of 1932, Johnson identified the zone that eventually became the Bankfield mine.

Toward the end of June, 1932, Johnson teamed up with Tony Oklend, a trapper with a cabin on Magnet Ck. Many years before, Oklend had chipped gold from a boulder on Kenogamisis L. Until Smith came along, no one had found anything else.

On July 5th, Johnson and Oklend staked the discovery claim that became the Little Long Lac mine. Sudbury Diamond Drill Co. took the initiative and began development. A staking rush ensued.

The Little Long Lac gold camp stretched from Wildgoose L. in the west to Kenogamisis L. in the east, the largest lake. The camp comprised an area just slightly larger than the townships of Ashmore, Errington, and Lindsley. As in the case of the Sturgeon River gold area, there was no road link to the rest of Canada. Most of the active prospecting occurred south of the railway line, which ran east-west, north of the largest lakes, Wildgoose and Kenogamisis.

The continental glaciers had deposited a thick mantle of gravel and sand. Extensive swamps and pothole lakes separated hills and ridges of glacial till. Bedrock exposures were small and scarce. Wildgoose L. drained westward into Lake Nipigon, which in turn drained eastward as part of the Atlantic watershed. Magnet L. and Kenogamisis L. drained eastward into the James Bay watershed.

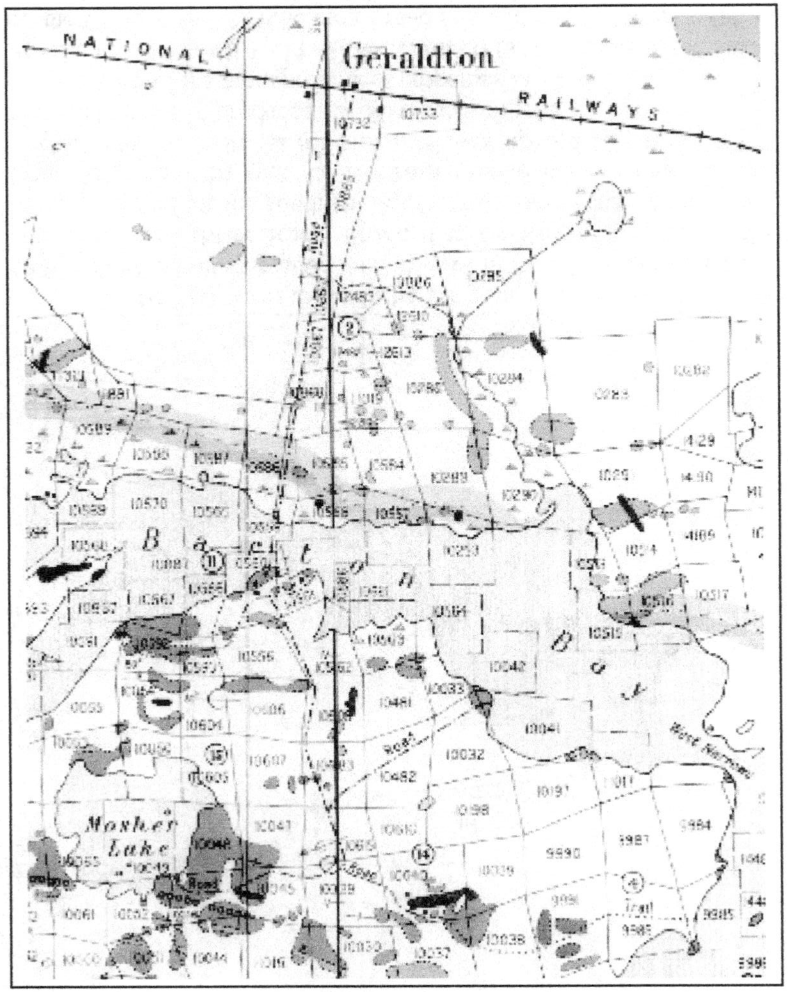

Section of Map 44d, "Little Long Lac Gold Area", ODM 1935

E.L. Bruce described the topography and geology of the Little Long Lac Gold Area in the ODM 44th Annual Report, Part III, 1935. A hill or ridge of rock followed the northern boundaries of both Ashmore and Errington Townships, so that the land sloped imperceptibly southward.

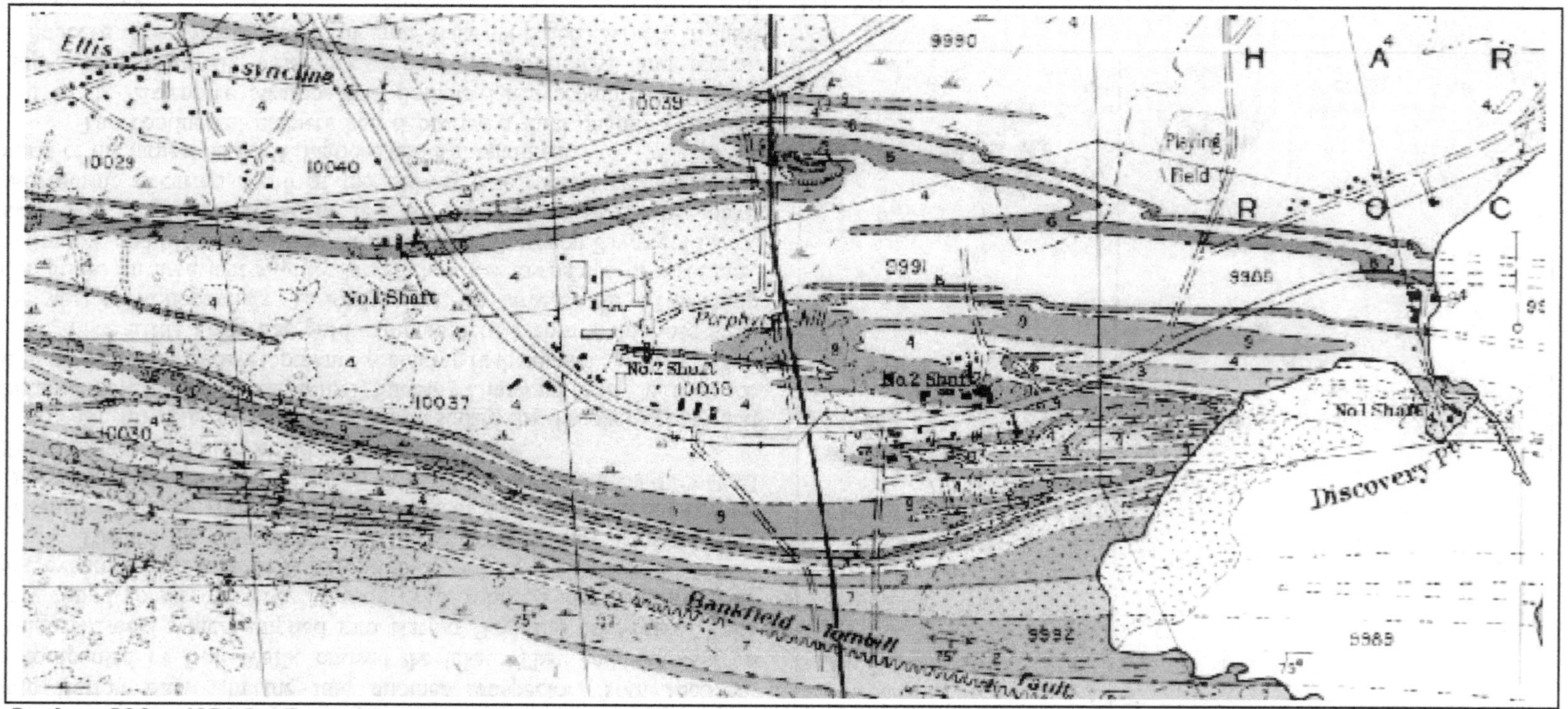

Section of Map 1951-2. "Township of Ashmore", ODM 1951

The terrain hosted large marshy and swampy areas broken occasionally by sandy and gravelly ridges or hills. Lakes were shallow, with very few rock exposures along the shores. Small, sluggish streams abounded. Spruce stands had good-sized trees.

Precambrian rocks underlay the glacial deposits, including highly altered lavas (Keewatin series), altered sediments (Timiskaming), granitic and dioritic intrusives, and diabase occurring in dikes.

Bruce identified gold as the area's only economic mineral. Miners had discovered gold-bearing veins only in the northern part of the east-west belt of Timiskaming sediments. The most favourable host rock was massive greywacké or arkose lying between two bands of iron formation (a description of the Little Long Lac mine structures).

The first producer in the Little Long Lac camp, **Little Long Lac Gold Mines, Limited**, poured its first brick in the fall of 1934. A bridge and motor road (Mine Road) linked it to Geraldton station two miles north. Located on the south shore of Barton Bay, it brought in power from Cameron Falls (via the Northern Empire mine) using the CNR right-of-way. In 1936 the seventh level was opened at a depth of 950 feet. In that year it produced 42,790 ounces of gold.

A list of the more prominent properties in 1936:

MacLeod-Cockshutt Gold Mines, Limited (south of Geraldton station on the CNR, with 3 ½-mile motor road link and hydro power) – surface work (SW), diamond drilling (DD), No. 1 shaft sunk to 172 feet, No. 2 shaft sunk to 532 feet

Hard Rock Gold Mines, Limited (adjoining the MacLeod-Cockshutt mine on the west, with Kenogamisis L. on the east) – SW, DD, No. 1

shaft on Discovery Point sunk to 475 feet, and a winze to 575 feet, No. 2 shaft on Porphyry Hill established a 150-foot level

Bankfield Consolidated Gold Mines, Limited (on Magnet L., 3-mile road link to CNR's Kenwell Siding in the north, 9 ½-mile motor road link to Geraldton) – SW, DD, shaft sunk to 525-foot level, 100-ton mill under construction

Tombill Gold Mines, Limited (adjoining the Bankfield mine on the east) – SW, DD, 417-foot shaft

Jellicoe Consolidated Gold Mines, Limited (adjoining the Tombill mine on the east) – SW, DD

Magnet Consolidated Mines (1936), Limited (adjoining the Bankfield mine on the west) – SW, DD, shaft-sinking reached 80 feet

Hutchison Lake Gold Mines, Limited (near Hutchison L., 4 miles north of Geraldton station, wagon road link) – SW, shaft-sinking reached depth of 75 feet

Elmos Gold Mines Limited (adjoining the Mosher Long Lac property and the MacLeod-Cockshutt mine on the east) – SW, DD

Mosher Long Lac Gold Mines Limited (adjoining the MacLeod-Cockshutt mine on the east) – SW, DD

Theresa Mine (strictly speaking, adjoining the Little Long Lac camp on the west, 6 miles south of Longlac) – SW, shaft-sinking begun, produced 13 oz. gold and 2 oz. silver in '35

In 1936 there were dozens of other properties in which owners had invested heavily in terms of time, resources, and money, for their faith never wavered that elsewhere in the Little Long Lac camp, there was an undiscovered star to equal the brilliance of the Little Long Lac mine.

Early Mines

Bankfield Mine

Chapter 1 : Year 2008

KODIAK
EXPLORATION LIMITED KXL: TSX.V

Looking west at Bush Lake Resort in the foreground and Pasha Lake Cabins in the background, November 15th

Activities of Kodiak Exploration Limited

NSD
NORTH STAR DRILLING LTD.

Muskeg Tours 2009

KODIAK EXPLORATION OPEN HOUSE

On Wednesday, March 5, Kodiak Exploration Limited held a public information session in Beardmore's Community Centre from 4:00 to 7:00 p.m.

As Larry Hillesland, project manager for Kodiak, explained, "The community can come in and see what we're doing and have planned". The open house is a requirement for the Advanced Exploration closure plan (AE) process mandated by the Mining Act.

Kodiak began exploration of its Hercules Project in August of 2006. The project encompasses about 6,000 hectares beginning around kilometre 31 on the Kinghorn Road, north of Highway 11. The exploration for precious metals has identified key gold zones – the Golden Mile, Yellow Brick Road, Marino, and 7 of 9.

There has been extensive trenching (stripping of overburden) and diamond drilling. Gold mineralization has been intercepted to 300 metres below the surface. Intercepts range from 1.06 to 25.24 grams per tonne (g/t). Some drill holes have encountered bonanza grades, including HR07-51 with 358.56 g/t over 3.6 metres in length. An troy ounce of gold is 31.1 grams.

The Kodiak Open House on March 5 was a busy venue. Presenters walked individuals through the wall charts and let them handle the core samples. All the while a PowerPoint presentation ran on a large screen with some very interesting video clips. People helped themselves to coffee and baked goods.

Doug W. Steele explained the role of TBT Engineering Consulting Group (TBTE), based in Thunder Bay. Steele, manger of environmental services, said that TBTE was hired by Kodiak to facilitate the AE process.

Tim Cano, TBTE biologist, indicated that an environmental baseline is to be established to help guide Kodiak's rehabilitation efforts. Part of his job involves identifying sensitive areas, such as eagle nests moose calving grounds, woodland caribou habitat, and fisheries values.

Aerial view of Kodiak's Golden Mile zone in late February, 2008. TBTE Engineering Photo.

Exploration by Kodiak on the Hercules Project has left several kilometers of shallow overburden trenches open to bedrock. A trench may be anywhere from 5 to 50 metres wide. Plans call for more than 60,000 square metres of trench to be rehabilitated this year.

Larry Hillesland (right), Kodiak project manager, makes his point as he gestures towards drill core exhibits. Looking on are, left to right, Scott Peterson, Amede Lafontaine, and Scott Shields.

Overburden (soil and organic matter) will be distributed in trenches to cover bedrock and to slope the sides of the trench. Much of the trench area is expected to recover through natural regeneration, though seeding and planting may be required in places. Other erosion controls may be implemented.

Part of the job of Scott Peterson, TBTE manager of geological services, involves groundwater and surface water assessments. He also explained that TBTE has 5 team members working on the AE process.

Hillesland, Kodiak project manager, pointed out that his company has an interest in claims covering about 1,400 square kilometers of the Beardmore-Geraldton gold belt. He was very pleased with the heavy turnout to the open house. After two hours, he had been involved in deep discussions with perhaps 20 people.

Hillesland spoke knowledgeably about the three types of drill core exhibits. The altered granodiorite represented the country rock or host rock. A sheared gabbro dyke is associated with the gold-bearing quartz vein. The quartz displayed large clumps of visible gold from hole number HR07-65.

He explained that the Advanced Exploration closure plan will allow Kodiak to extract bulk samples from representative sections of the Golden Mile zone. Rock extracted from up to 5 different places could amount to 1,560 tonnes per sample. If, for example, a sample is 10 metres deep and the vein 2 metres wide, and rock is extracted for a total of 30 metres along strike, the yield will be 600 cubic metres of mineralized rock. At 2.65 tonnes per cubic metre, the bulk sample would amount to 1,560 tonnes.

Mark O'Brien was on hand representing the Ministry of Northern Development and Mines (MNDM). O'Brien, mineral development and consultant, is MNDM's regulator for the 2008 closure plan. The plan will likely be finalized in April. Kodiak will be required to give "financial assurance", which means that it will place funds in trust with MNDM to cover the cost of rehabilitation. As rehabilitation work is completed, the funds can be released.

O'Brien explained that the 2008 closure plan can be extended to 2009 by an amendment. By 2010 Kodiak could be entering development and production stage, with expanded requirements under the Mining Act.

A number of other government ministries are involved in the AE process, for work permits, waste management, environmental assessment, protection of fish and wildlife, and permission to take water.

Currently there are three drill rigs on the Hercules Project and a non-technical work force of 18, a quarter of whom are First Nations members. In 2008 Kodiak proposes up to 20,000 square metres of new trenching and more than 40,000 metres in core drilling.

Line cutting for geophysical studies will involve about 20,000 metres (20 kilometres), and trench reclamation, under the closure plan, will involve more than 60,000 square metres.

In 2009 Kodiak is proposing more trenching and more trench reclamation efforts. The drilling program will continue, and if the results of the 2008 exploration warrant it, underground exploration.

By about 6:00 p.m., more than 50 people had signed the guest book and more were streaming in. Most people identified Beardmore or Geraldton as their place of residence. Several people, of course, had not signed in for whatever reason. In the end, presenters estimated attendance at just over 80.

On March 5, gold was trading on the market at about $990 Canadian per ounce.

- Times Star

THERE'S GOLD BEYOND BLUEBERRY HILL

In its search for gold, Kodiak Exploration Limited is leaving no stone unturned. Literally. It is examining ground that prospectors abandoned more than half a century ago.

In the 1930's and '40's, prospectors looked seriously at the Portage Shear Zone Fault, which has an east-west trend that passes through Geraldton's urban area, north of Barton Bay on Kenogamisis Lake. But all the action seemed to be happening on the Bankfield-Tombill Fault, which trends east-west, south of Barton Bay. That fault system yielded nine producing mines.

This past winter Stephen Roach, senior consulting geologist for Kodiak, researched properties west of Geraldton. His findings led him to recommend exploration in this area, and the results of an airborne survey run on the whole Greenstone region this spring supported his findings.

In the spring the Municipality of Greenstone agreed to option a block of claims to Kodiak, a block Kodiak calls its Geraldton Kenogamisis Property. On the morning of July 29, the writer accompanied Stephen Roach and Gerry White to one of the two active exploration targets on this property.

Taking the unpaved road that runs past the Geraldton Medical Clinic and behind the water tower, the group traveled in pickup trucks to the area known locally as Blueberry Hill. According to Herb Byerley, the name refers specifically to a rocky outcrop behind the municipal trailer court and the cemetery, also known locally as the Junior Rockies. The long-time resident also recalled that a company called Lafayette had a shaft north of Barton Bay, and he recalled seeing cabins on the north shore past the big island. Records indicate that the shaft was sunk on the "east zone" of the Geraldton Kenogamisis Property.

The party drove past Blueberry Hill on a single-lane road that Kodiak had recently upgraded. Past the hydro line corridor, the party took a trail recently gouged out of the bush by a backhoe. In a swampy section the party encountered incredibly rough corduroy, so they resorted to walking the last 800 metres. The weather was cool and cloudy, with few flies, and the forecast for the afternoon was a thunderstorm.

Just under 4 kilometres from the water tower, the "west zone", on Claim No. TB 10943, has been recently stripped of bush and soil. An area of bedrock, about a football field in extent, has been exposed in a northeasterly direction, with a branch stretching westward to expose the No. 1 vein.

Roach pointed to the vein, trending east-west, and smiled: "I call it the No. 1 to Geraldton." He was alluding to the recent Western movie starring Russell Crowe, "3:10 to Yuma". For lack of an official name, Roach called the area the No. 1 Vein.

About 600 metres further east through the bush lies the "east zone", officially dubbed the Goldstrike Zone. Nearby is also the Drexel Zone. One accesses these zones by traveling a rough trail south along the hydro corridor. The terms "east zone" and "west zone" were applied by miners back in the '30's and '40's.

Back at the No. 1 Vein, two geotechs were washing down the exposed northeast-trending No. 6 vein with a hose. Eric Deroy and Richard Metansinine also operated water pumps and cut channel samples. Deroy had been working on the Hercules Project until Roach requested his assistance on the Geraldton Kenogamisis Property, closer to Deroy's home in Geraldton. Metansinine, from Jellicoe, was temporarily filling in for Andy Brown, another geotech.

Further south in the clearing, where the northeast-trending No. 3 vein intersected the east-trending No. 1 vein, prospector Phil Houghton (from Beardmore) was examining channel samples. Terry Halverson, a prospector from White River, and Greg McKay, a geology student from Ottawa, assisted Houghton. Said Houghton, "We're just doing the samples and waiting for the results to come back [from assays]. This is the best-looking arseno so far."

Gerry White, district geologist for the Ministry of Northern Development and Mines, was visiting the property for the first time. The work of stripping had begun about mid June. Steve Leduchowski Trucking Inc. had stripped about 2700 square metres. This particular zone exhibited no visible gold, but White was visibly impressed by the "stockwork", the network of gold-bearing veins and country (host) rock. Several times he exclaimed over the "indicator" minerals, especially the arsenopyrite.

Roach pointed out to White a rare mineral called scorodite, not easy to spot. Scorodite is a greenish yellow mineral formed by the oxidation of arsenopyrite. Arsenopyrite (FeAsS) is a mineral frequently associated with gold. Its distinctive crystals, metallic luster, and colour ranging from brassy white to grey make it a favourite among collectors. When crushed, it emits a garlicky odour.

All three vein systems in the clearing had been trenched in the '30's. Further to the south, in the bush, Roach pointed out some flagging tape on trees that indicated the locations of two more known veins, No. 10 and No. 11, still to be exposed.

The old trenches in the bedrock ranged from a few centimetres to a metre in width. Roach stated that the trenches on the No. 1 vein ranged from 5 cm to 1.2 m wide, and averaged .25 m in width. Occasionally a pit, up to 2 metres deep and up to as many metres wide, had been sunk by the oldtime miners.

White explained, "They were constantly panning this stuff, taking it out and grinding it with mortar and pestle." If they found good colour, they would dig deeper.

Roach had been surprised by the scarcity of "fly rock", the rocky debris blasted from the trenches. The oldtimers would have drilled holes

Gerry White of MNDM and Stephen Roach of Kodiak stand on either side of the trench for the No. 1 vein. Prospectors dug the trench in the 1930's. View looking west.

by handsteel, packed in explosives, and excavated the narrow trenches and the pits. He concluded that they had been "mining" the fly rock. In the early stages of exploration, there would not have been any decent road to the find. "They mined the stuff," said Roach, "and packed the stuff out by horse or something in the '30's."

From his research of old records, Roach had determined that the property did have some interesting gold values, enough to prompt the oldtimers to initiate drilling programs. However, the results from shallow holes did not justify more intensive exploration in those days. Especially when their colleagues were making spectacular finds and establishing producing mines on the Bankfield-Tombill Fault.

Kodiak has sent samples of fly rock to an assay lab, but Roach said he would not be permitted to divulge the results until Kodiak makes an official announcement. On August 1, Kodiak released a comprehensive report on results of its Hercules Project and some information on other properties in the Greenstone District. The only mention of the "west zone", still without an official name, is as follows:
"A second zone of strong mineralization located 500 metres west of the Goldstrike zone has also been discovered. These structures are characterized by quartz veining with arsenopyrite over broad widths where historic grab samples have graded up to 129.5 gpt [grams per tonne] gold (3.777 opt). Additional stripping and channel sampling is ongoing; drilling has commenced."

On July 29, Roach said he was expecting a drill to arrive shortly from the Hercules Project.

The No. 1 vein had been stripped for a distance of 120 metres. In some places in the narrow trench, arsenopyrite dominated the wall (host) rock.

White was also fascinated by other promising indicators. The host rock was mafic volcanics (with pillow lava structure) and altered sediments. The pillow lavas exhibited evidence of stretching. The host rock was folded, sheared, fractured, cross-faulted – a jumble of interesting structures. "If you see a mix of all kinds of rock in one zone," said White, "that's an area you want to look at."

Roach had mentioned earlier that Kodiak has been exploring an area further west of Geraldton, called the West Geraldton Property. It is a block of claims south of the railway track, east of the Kenwall Siding Road, and north of Magnet Lake. Kodiak prospectors found old trenches on the Portage Shear Zone Fault. Kodiak cut a grid of north-south lines through the bush, and carried out a ground geophysical survey over five weeks in May and June. The area is quite swampy, and would lend itself to a winter drilling program.

With rain clouds moving in, the party abandoned the plan to proceed to the Goldstrike Zone. Times Star will report on this zone at a later date. The next article in this series will survey the history of the Geraldton Kenogamisis Property.

- Times Star

GOLDSTRIKE : AN ARCHEAN GEOLOGIC PARK

On Thursday morning, August 28, the writer and three geologists drove the Blueberry Hill Road to examine the Goldstrike Zone of the Geraldton Kenogamisis Property. Kodiak Exploration Limited optioned the gold property this spring from the Municipality of Greenstone.

On the west side of the hydro line, the two pickup trucks followed a very rough trail recently gouged out of the bush for about 400 metres. Parking the trucks, the party struck out on foot, following an ATV trail on the west side of the power corridor. For a distance, the backhoe had churned up a trail on the east side of the corridor. Then the backhoe had crossed over and continued the access trail through the bush, having bypassed a hopelessly swampy area.

On the last hill before the Goldstrike Zone, the party left the ATV trail, cut through the bush for a few metres, and joined the access trail. One point four kilometers from Blueberry Road, the party emerged in the clearing where the target area has been stripped of overburden. The Goldstrike vein trends generally west from the hydro line for about 300 metres (from a compass bearing beginning at approximately 255 degrees and swinging northwest to 290 degrees).

Back in the thirties, when Lafayette Long Lac Gold Mines Ltd. began exploring this target (which they called simply "the east zone"), they started at bedrock exposures, and traced the vein by sinking overburden pits. The vein obviously yielded good gold values, for no visible gold remains today, and some shallow pits sunk into the bedrock have completely obliterated the vein rock. There are no official records of Lafayette's surface sample values. The results of Kodiak's channel samples have not been released.

Stephen Roach, Kodiak geologist, points to the rust-streaked Goldstrike vein where it descends into the pit created by removing the overburden. Two other geologists take notes, Trevor Bremner and Gerry White.

The Goldstrike quartz vein is narrow, from a few centimeters to perhaps half a metre wide. At the east end, Kodiak has excavated an overburden pit two and three metres deep where the east-trending vein seemed to end on surface. The backhoe exposed the vein to a width of 2.5 metres. Stephen Roach, senior consulting geologist for Kodiak, commented, "It's a typical pinch and swell type of structure."

Kodiak has stripped the overburden westward, uncovering basins in the bedrock that the oldtimers had never explored. One of the party, Trevor Bremner, a visiting consulting geologist for Kodiak, expressed amazement at Kodiak's work, the bedrock being washed down and as clean as a dining table. "This is a man-made outcrop and it's so neat and tidy, " he said. "Stephen has opened a whole new world of geology we can look at." He added that the exposure provides so much more information than drill core can.

The rusty drill casing from the historic #2 Hole projects from the ground.

Bremner also stated that he was surprised to see a promising prospect with the infrastructure already in place. He was referring to the road system, the adjacent hydro line, and the nearby community.

During the entire field trip, the intermittent rain proved a boon, wetting down the bedrock and veins, and providing superior views of the minerals and structures.

The Goldstrike quartz vein and altered host rock exhibited a rusty colour, emanating from the iron carbonates (siderite/ankerite). Other vein minerals included calcite (calcium carbonate), dolomite (calcium magnesium carbonate), fuchsite (a chromium-rich mica, emerald green), tourmaline (black fibrous crystals), arsenopyrite, pyrite, pyrrhotite, and occasional chalcopyrite.

The field party next examined some historic drill casings near the Bakery Zone (a short vein system named by Roach), which runs on the south side of Goldstrike. When Kodiak had sent in a prospecting party to rediscover the historic east zone, they had started further south and worked their way north. When they found three casings, they knew they were in the right area. They had also learned that the overburden pits had been concealed by vegetation and soil.

At historic Hole #2, the metre-and-a-half casing projected out of the ground in a southerly direction, disguised by rust and lichens so that it resembled a stick or branch. Some values from historic holes have been preserved, and we know that 4.73 feet (1.44 m) of Hole #2 assayed at $335. per ton, which, at $35 per ounce, translates into approximately 9.57 ounces per short ton (328 grams per metric tonne). This value stood head and shoulders above the other historic drilling results, which were on the whole disappointing. Kodiak recently drilled a hole to intersect the Hole #2 underground target from the opposite direction, from north to south. The assays of Kodiak's 11-hole program, a total of 1570 metres, have not yet come back.

As a matter of interest, old records show that Lafayette sank a 35-foot shaft in this vicinity, a shaft which Kodiak's prospectors have not located yet. This should be a caution to tourists who want to poke around the site.

Proceeding west along the Goldstrike vein, the field party encountered a diabase dike, 20 metres wide, intersecting the vein. Next they encountered a previously unknown shear system, called the Drexel Zone by Roach. The Drexel Zone loops for 140 metres to the north of the Goldstrike vein before merging with it again.

The country rocks that host Goldstrike/Drexel are an amazing geological phenomenon. The host rocks are iron- and magnesium-rich basaltic flows that show well-developed pillow textures. Pillow lavas form when active volcanic flows come into contact with water. These pillows (Think of lobes or blisters of rock) formed in a shallow marine environment, i.e., in less than 500 metres of water.

The Goldstrike vein is intersected by a black diabase dike, evidence of continued magmatic activity.

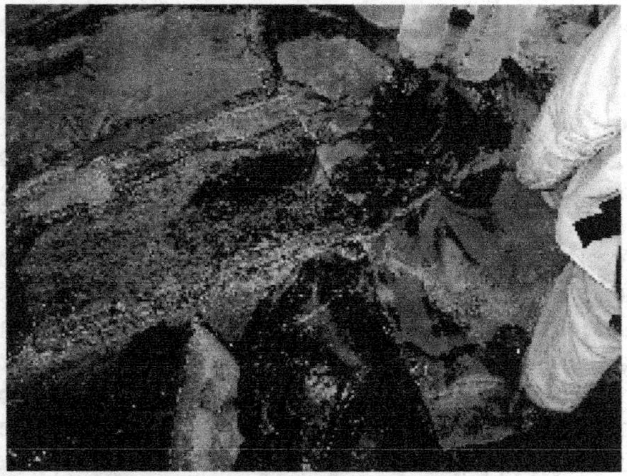

Pillows can range from a few millimeters to several metres wide, but in the case of the Goldstrike/Drexel host rocks, the well-defined pillows are twenty to thirty centimetres wide. They are all flat, each pillow surrounded by a distinguishable border, the bedrock having been eroded by glacial action.

There were also beautiful examples of "stretched" or altered pillows, where only the sharp eye can make out the occasional fragment of border. In some spots, quartz filled small cavities in the borders, producing the visual effect of a charcoal gray sculpture punctuated with glowing white enamels.

Stephen Roach said that Kodiak is primarily interested in the possibilities of gold deposits in the host rocks of these zones. The gold-bearing quartz veins are simply important indicators. The drilling program results will be able to provide valuable information.

In the Drexel Zone, some surfaces exhibited smooth and wavy or channel-like contours, as well as striae (long scratch lines), evidence of erosion by continental glaciers.

The three geologists marveled at the geological phenomena that come together on this site. Someone suggested the descriptive term "geologic park". Aside from the potential for gold deposits, the site would be of enormous scientific interest to academics and to students of prospecting. The oldtime miners had been completely oblivious of this zone.

The Goldstrike Zone lies due east of the No. 1 vein system (the historic "west zone"), visited by the writer on July 29. Roach said a drilling program would begin in early September in the No. 1 Zone area (The target area still lacks an official name). He mentioned that the No. 10 and No. 11 veins of that zone have been stripped, and that the No. 6 has been exposed along an arc in such a way that he called the zone Boomerang.

Roach mentioned that evidences of prospecting have also been found between the historic east and west zones. Meanwhile, geologists were completing the mapping of the surface of the Goldstrike target area.

If Kodiak Exploration Limited does not find a mine on Goldstrike, perhaps the community can interest Steven Spielberg in a prequel to his *Jurassic Park*, filmed on location in Greenstone's Archean Park.

- Times Star

The pillow lava flow has been "stretched", and quartz has infiltrated the cavities.

HISTORY OF GERALDTON KENOGAMISIS PROPERTY*
Part I – July of 1934

In July of 1934, there was no town called Geraldton. There was a small clearing in the bush called Geraldton Station, and a rough trail running south to Barton Bay of Kenogamisis Lake. If anyone in the world had heard of the place, they knew it as the Little Long Lac gold camp.

In July of 1934, the Little Long Lac mining camp may have been the most exciting place in Canada.

In July of 1934, Hard Rock Gold Mines Ltd. was still diamond drilling its property near the southwest arm of the lake. Next door, Macleod-Cockshutt Gold Mines Ltd. was sinking a three-compartment exploration shaft. Further afield, at Magnet Lake, Bankfield Gold Mines Ltd. was reaching the conclusion that it needed a shaft.

But none of them was the lodestar that was drawing thousands of miners and prospectors and entrepreneurs and just plain curiosity-seekers to the goldfields. That star was Little Long Lac Gold Mines Ltd. In July of 1934, a shaft had already been sunk on the south shore of Barton Bay. A headframe had been erected. A huge milling complex was under construction. The company had extended the hydro power line from the Northern Empire mine in Beardmore, which in turn drew power from Cameron Falls on the Nipigon River. Communities were growing up around the mine, communities we now know as Little Long Lac Townsite and Rosedale Point.

In July of 1934, there was no town of Geraldton. The place that held promise of becoming a town was Hardrock Station, a few miles to the east on the CNR railway. Hardrock sat on the shore where the railway skirted the lake. Hardrock was the jumping-off point for the goldfields.

Hardrock had stores, boarding houses, eating places, a poolroom, a bank, offices for a broker, an assayer, and a lawyer, and for mining companies. Freight cars loaded with tons of machinery and supplies jammed the railway siding. Every canoe, boat, and scow was employed to move people and goods to the mines and to the prospects around Kenogamisis.

*** Kodiak has renamed it the West Geraldton Property. The No. 1 Vein showing is now the Golden Boomerang Zone.**

Aerial view of Little Long Lac gold mine under construction, looking generally north across Barton Bay. Circa the summer of 1934. Note the hydro line coming down to the north shore beside the trail.

A writer from *Gold* magazine described a trip from Hardrock to Little Long Lac mine in July of 1934. "On every headland and wind-swept point freshly cut stakes gleam sharply in the sunlight against the verdant green of the bush. White tents or cabins of freshly-peeled logs interrupt the marching ranks of pine and birch at intervals along the water's edge . . . . The smoke from camp fires pillar and away with every passing breeze. Prospectors, Indians, surveyors, all wave a cheery greeting as you pass."

In July of 1934 a trickle of reports began about a very small but a possibly interesting prospect on the north shore of Barton Bay. Other prospects received more coverage: Jellicoe (a prospect, not the village), Longlac Lagoon, Mosher, Longacre Long Lac, Roche Long Lac, for example.

The very small but possibly interesting prospect was being explored by Lafayette Long Lac Gold Mines Ltd.

Today, Kodiak Exploration Ltd. calls that prospect its Geraldton Kenogamisis Property (Editor's Note: Now part of West Geraldton Property).

Part 2 – June to December, 1934

This year Kodiak Exploration Limited has been conducting a mineral exploration of the Portage Shear Zone fault, the most intensive since the 1930's. The Geraldton Kenogamisis Property is located immediately west of the community of Geraldton, north of Barton Bay.

Back in 1934, the first inkling the general public had that this property might be important came in a brief news item dated June 21 in *The Northern Miner* (*TNM*): "A number of other properties are being

Two geologists, Stephen Roach of Kodiak, and Gerry White of MNDM, stand on massive volcanics that host the Goldstrike quartz carbonate vein before them. The vein dips into the overburden pit created by a backhoe. Beyond them, to the east, is the hydro line corridor. Behind them, Lafayette had located Holes #1, 2, and 3, drilling from south to north.

explored. Lafayette Gold Mines, to the north of Little Long Lac mine, is opening up a vein in which gold values have been found."

On July 5, the *TNM* reported that Lafayette had brought in a portable gasoline-driven compressor to open up surface veins. We do not know with certainty what work the compressor performed. However, in its August edition, *Gold Magazine* reported "the cores of a light diamond drill range from $1.80 to $30 per ton" (Refer to sidebar, **Gold Values**). Throughout July, the *Fort William Daily Times-Journal* (*DTJ*) provided a stream of brief items. On July 6, the *DTJ* reported an on-site engineer as saying "considerable free gold is showing up in the number three vein", while nearby a new vein, 12 inches wide, was "well mineralized". Furthermore, "The number six vein has widened out to four feet of quartz, all mineralized."

On July 11, the *DTJ* reported 12 major veins in a shear zone 800 feet wide, with trenching in an east-west direction. These clues, and others, suggest the target area was what the company would eventually call its east zone. Three days later, Lafayette reported acquiring 5 new claims in the northwest section to create a total holding of 23 claims (later amended to 25 claims). It appears that work was now started on what the company later called its west zone. On July 31, the term "east zone" first occurred in print.

Meanwhile, the *TNM* was issuing brief progress reports. The *TNM* noted extraordinary progress on veins discovered and work performed. On July 19, Layfayette's president reported more than 5,000 feet of trenching along a length of 1,500 feet and across a width of 600 feet (compared to the *DTJ*'s report of 800 feet). He reported 5 veins (*DTJ* : 12 veins) ranging up to 5 feet wide, and for the first time, there is a mention of test pits, 6 in number, from 6 to 8 feet deep.

The early miners were no strangers to boosterism, and the regional newspapers and the stringers for the publications with province-wide, even national, distribution, knew what their sources (the mining industry) and their readers wanted: News with a capital "N". In almost every case, no journalist actually visited the prospect. And there is also the very real possibility that reporters who were unfamiliar with mining operations and mining lingo did not represent the facts accurately. This problem persists today.

The evidence on the ground today in the historic east zone, now Kodiak's Goldstrike Zone, suggests a different story. To access the veins, there may have been deep trenches and pits in the overburden, for the outcrop areas were small and few in number. The veins themselves averaged 0.20 metres (6.5 inches) in width. Trenches in the veins ranged from 0.50 to 1 metre in depth, with occasional shallow pits up to 2 metres (6.5 feet) deep. There is no way that early miners carried out the extensive stripping operations that Kodiak conducts today. The reports of a zone 600 or 800 feet wide and 1,500 feet long may have led many a reader astray. Certainly the oldtimers sank test pits across a good width and along strike, but the pits were few and far between. If they encountered a vein, they followed up the vein until it pinched out, or dipped too deep.

On August 2, the *TNM* reported: "On what is referred to as the west zone six veins have been uncovered. Vein widths vary from 18 inches to five feet." On August 23, an item stated that Lafayette was proceeding with channel and bulk sampling, that the company had let a diamond drilling contract, and that it was completing permanent camps. According to news reports that summer, the work force numbered between 20 and 29 at different times.

Another news item in August bears examination. On August 14, the *DTJ* stated, "The east discovery is improving and has now been traced for 700 feet with free gold still in the quartz averaging 8 feet in width." The source was the on-site engineer in charge of operations. Compare to the *DTJ* statement from the company president on July 19 about 5,000 feet of trenching over 1,500 feet. However, the "averaging 8 feet in width" was a nice touch. The maximum vein width confirmed by Kodiak in the Goldstrike Zone is about 2.5 metres (8.25 feet) in a single spot, with the "average" width being 0.20 metre (6.5 inches). However, the engineer was likely referring to the shear zone (the altered rock) and not the quartz vein. The shear actually varies considerably in width, to a maximum of about 4 metres (13 feet).

News for the remainder of the year is sparse. There is a single, unsourced report in the *TNM* dated November 22. It is actually an extract from the Annual Report of the Department of Mines, and it describes the geology: "The rocks exposed on the claims are volcanics, mainly lavas but with some tuffaceous beds. Diorite intrudes the volcanics . . . . On claim T.B. 11,321 [the east zone] . . . [q]uartz veins up to one inch in width cut through the [eight-foot] shear zone . . . . [G]old appears in visible quantities in several places . . . . On T.B. 10,943 [the west zone] . . . . [t]he zone has been traced for 700 feet . . . Parallel to it is another zone . . . Low values in gold are obtained on assay." Minerals mentioned are pyrite, chalcopyrite, carbonate, ankerite, and arsenopyrite.

The *Canadian Mines Handbook* for that period stated that Lafayette Long Lac Gold Mines Limited was incorporated in 1934 with 3,000,000 at $1 par value, with 11 claims and options on 9 claims, all in the Little Long Lac area. The foregoing was likely the incorporation or prospectus information. The next annual edition listed Lafayette with 25 claims.

There is a sketch map of Claim T.B. 11321 (the historic east zone, and Kodiak's Goldstrike Zone) which has no date and no author. It shows isolated trenching areas over a length of about 1,000 feet in a zone with a maximum width of 200 feet. It shows the location of diamond drill holes #1, 2, and 3, and possibly a fourth hole, all concentrated in the eastern sector, not far from the hydro line. In the neat script of a draftsman or engineer, it lists gold values for a total of 37 surface samples (locations marked on the map) and 6 (or 7) assays from the drill holes. Refer to the sidebar, **Gold Values**.

The values were calculated on $20 per oz. This fact suggests a map date between late 1934 and early 1935.

Exploration carried on briskly in 1935.

Gold Values

In 1934 Canada's gold was priced around $20 per troy ounce. Therefore, values ranging from $1.80 to $30 per ton (1 short ton = 2,000 pounds) translate into 0.09 oz. and 1.5 oz. per ton. One metric tonne is 1,000 kilograms, or 2,200 pounds. Therefore, the values convert to 3.06 grams and 51.43 grams per tonne.

At that time, Canada, no longer on the gold standard, still applied its old statutory price of gold, namely, $20.67 per ounce. By an Act of Parliament in July of 1935, the value changed to the prevailing world standard of $35. per ounce.

The sketch map for Claim T.B. 11321 recorded 19 of the 37 surface samples assaying at less than a dollar per short ton. Only 5 assayed at $5.00 or more, by far the best being $75.60.

Hole #1 returned assays of $4.40, $14.00, and "free gold"; Hole #2, $36.80; and Hole #3, $4.40 and $0.20. However, these values don't correspond to values reported by consulting engineer W.W. Beaton for these hole numbers in July of 1935, for the latter were a lot higher. The mystery is explained by a report in the *TNM* on May 2, 1935, when Lafayette was starting a new drilling program: "The light diamond drill used in putting down holes Nos. 1, 2 and part of No. 3 has been replaced by a heavier drill." The heavier drill could have sunk deeper holes, or replicated the original ones.

Part 3 – 1935-6

In July of 1935, consulting mining engineer W.W. Beaton recommended a program of work on the "Geraldton Property of the Lafayette Long Lac Gold Mines Limited". The July 19th report listed 25 claims, including Claim No.'s TB 10943 and TB 11321.

The topography was described as "one of very low relief, the greater part of which is swamp land. Spaced at irregular intervals and with varying area are small rounded knolls in the swamps. These knolls are bedrock covered with shallow overburden and moss and are the only places on which surface work may be performed."

Beaton stated that all work to date had been confined to the claims cited above. On TB 11321 (Kodiak's Goldstrike Zone) a shear zone had been traced for 400 feet, with a width of 9 feet. Lafayette had drilled 9 holes along this shear, only the first two of which had economic values. Beaton reported "several" parallel shear zones being trenched and blasted for sampling.

As for the "west zone" on TB 10943, the report stated "these veins were very narrow, the average width being about 12". Test pits were sunk at regular intervals along three of the veins and, on sinking through the zone of oxidation, the veins increased rapidly in width to a maximum of 7'."

The three veins of interest were No. 1 (exposed for 400 feet), No. 3 (300 feet), and No. 12 (100 feet). "These veins are heavily mineralized with arsenopyrite, pyrite, pyrrhotite and chalcopyrite." Gold values were low.

Beaton recommended drilling 4 holes on the west zone, some vertical holes on Claim TB 11222 (later described as the "south claim"), and trenching and blasting at regular intervals along the shears of the east zone.

Earlier in 1935, *The Northern Miner* (*TNM*) had reported on the Lafayette property. On March 14, the *TNM* stated that "camps capable of housing 26 men have been erected and are fully equipped". The Beaton report in July listed 6 buildings: "frame" cook camp, frame bunkhouse, frame storehouse, log bunkhouse ("two stories"), log office, and log residence. Beaton added, "Sufficient tools are on the job for a crew of 15 men." A report in the *NM*, July 11, stated "a crew of 11 men is presently engaged at the property".

Beaton's July report confirmed that the main access to the property remained Barton Bay. "The Company owns one 9 H.P. heavy duty Johnson outboard motors [sic] and one 3 H.P. Viking outboard motor. Also one large freighting canoe 20' long and one flat bottomed boat 20' long." There is no mention of road access. A map prepared by M.R. Brown in 1945 shows 3 camp buildings on the north shore of Barton Bay, and 2 camp buildings immediately south of the east-west line of drill holes on Claim TB 11321 (Kodiak's Goldstrike Zone).

This drill hole location on the Goldstrike Zone is known as a collar. The casing projects from the ground. A cap prevents entry of foreign material. Vital information is printed on the stick: COLLAR GK-08-11 / Dip: -50 / Depth 100 m.

A *TNM* report dated March 28 stated, "[T]he camps have been reopened and a road cut in preparation for diamond drilling." Sudbury Diamond Drilling (co-developer of the Little Long Lac mine) had the contract, and had drilled the first hole earlier in the week. "The first three holes have been spotted along the vein exposed in the east zone for a length of 130 ft. . . ."

On May 23, *TNM* reported a second diamond drilling contract let for the west zone. Beaton's report in July recommended that 4 holes be drilled on TB 10943, but there was no mention of an ongoing drilling program on that property. However, a *TNM* report dated August 1 stated that Lafayette had just begun the third drill hole on the west zone (TB 10943), and that a drilling program was proceeding, with Hole #13 just completed. This report confirms other evidence that the company sank the first 11 holes on the east zone, and began Hole #12 on the west zone.

On September 12, *TNM* reported that Lafayette had started Hole #20 on its south claim (TB 11222, a move recommended by Beaton in his report), and had just completed 5 holes on the No. 11 vein of the west zone. On October 31, *TNM* reported that the company had ceased field operations, and that "the work to date has failed to detail any definite ore occurrences."

Still, on November 14, *TNM* reported Lafayette planned to sink a hundred-foot shaft on the east zone, and to carry out underground work to investigate the high grade intersected by the #2 Hole.

On December 12, *TNM* reported the shaft had been sunk 30 feet by hand steel. This was wishful thinking, because on December 26 *TNM* reported that the 4 by 7-foot shaft had reached a depth of "over 20 feet" the week before. In addition, the target was now an intersection of the No. 1 Hole.

"The shaft is reported to start in the diorite-schist contact, which appears to dip vertically, immediately over the N— [type obliterated] hole intersection . . . The shaft is going down by handsteel and windlass operation with five men employed."

The Annual Report of the Department of Mines confirmed a two-compartment shaft started in December and sunk to a depth of 30 feet.

Another Department document listed "estimated employment" during 1935 as 10 per month from May to October, 6 for November, and 8 for December.

Another report by W.W. Beaton, dated May 4, 1936, confirmed that of 20 holes Lafayette drilled in 1935, the east zone had the first 9 holes along the shear (Another document assigned Holes #10 and 11 to the east zone as well, with no significant values), the west zone Holes #12 to 19 (with the last 5 holes on the No. 11 vein), and the south claim (TB 11222), the #20 Hole.

The *Canadian Mines Handbook* for that year reported for Lafayette that there had been "considerable" surface exploration, approximately 6,000 feet of diamond drilling, and a "prospect shaft being sunk to investigate high grade d d intersections".

Early in 1936, work on the shaft ceased. An item in *TNM* on February 13 reads in part: "[O]perations have been temporarily suspended at the property. The small plant which was being utilized for shaft sinking was not capable of coping with the water flow, it is understood, and after having reached a depth of 35 ft. in the shaft, work was discontinued."

There remains at this time the mystery of this shaft's location. The news item states the shaft was intended to investigate "two high grade intersections" (i.e., Holes #1 and 2). M.R. Brown's 1945 map identifies a "Pit" [another name for a shallow shaft] just east and north of Hole #1.

In 1936 another company would get involved in this property.

GOLD ASSAYS

A report in *The Northern Miner* dated June 27 provides some insight into the assay process. The Ontario Securities Commission investigated rumours that reports of high grades in Lafayette's drill cores were suspect. The rumours referred to Holes No. 1 and 2 of the east zone.

Lafayette had split the cores in two, lengthwise, with one half being sent to the assay lab (possibly in Haileybury). The Commission directed the company to submit the other half of the cores. The Commission concluded: "These were assayed by the Department of Mines and practically the same results were secured as published by the company as the result of the assays. There was apparently no justification for the complaint."

Inside Kodiak's core shack (behind the Park Bay Motel), Reg Seyler, geologist, logs core on a laptop.

LOGS OF HISTORICAL HOLES

Two logs of drill core of Marquette Long Lac Gold Mines Ltd. have survived. The following are extracts from the 1936 drilling program:

Hole M4 (intersections expressed in feet)**:** Angle at collar 60 degrees.

0.0 – 20.0 Casing.

20.0 – 23.0 Graywacke, altered to chlorite schist, fine grain dark green colour, somewhat silicified.

67.9 – 150.5 Arkose, medium grain, light colour, fair amount of fine to medium grain disseminated pyrite. From 70' the section is heavily carbonated. At 128' it becomes much coarser. Occasional small quartz stringers with pyrite and chalcopyrite.

222.0 End of hole.

Hole M5: Angle at collar 45 degrees.

0.0 – 25.0 Casing

25.0 – 32.0 Iron formation, becomes leaner towards the end of the section.

121.5 – 148.4 Conglomerate, pebbles sheared and drawn out. From 127-128.4, carbonated zone with considerable fine pyrite. From 137.0-138.2 numerous small quartz-carbonate stringers with some pyrite.

173.0 Hole not completed.

Some of the results even an amateur can interpret. There was no visible gold. "Hole not completed" suggests that the miner in charge gave up on the unpromising Hole M5. He moved on to Hole M6.

But that's not what happened. A record has survived giving the total length of Hole M5 as 309 feet. Sometimes a perfectly reasonable assumption is trumped by a fact.

Part IV – 1936 - 1943

In January of 1936, Geraldton was little more than a clearing in the bush, straddling the CNR tracks. A plan of subdivision showed two streets on either side of Main Street for a distance of just over three blocks, and three avenues north of the tracks.

Main Street boasted a few fledgling businesses, and the rest of the townsite, a scattering of homes.

The road that ran south to Barton Bay, called Mine Road, had not yet acquired its current name, Main Street. On the north shore of Barton Bay was another plan of subdivision called Johnsonville (later called Jonesville). A very long bridge spanned the Bay (There was no causeway) in order to access the operating mine of Little Long Lac, and the promising prospects of MacLeod-Cockshutt Gold Mines Ltd. and Hard Rock Gold Mines Ltd., and their respective townsites.

It might have been the incredible success of the Little Long Lac mine that encouraged Lafayette Long Lac Gold Mines Ltd. to keep the faith. The words "Long Lac" in Lafayette's name were no accident. But after two years of exploration, the surface work and the 4,975.4 ft.* of drill core had been less than encouraging.

What miners call a "trail" can range from a footpath to a motor road. On July 29, 2008 two geologists walk the new trail accessing the old Marquette west zone of Claim TB 10943. The stretch of corduroy, which only ATV's and heavy machinery can cross, later gives way to a perfectly good road.

Yet something was happening behind the scenes. An item in *The Northern Miner (TNM)* on May 14 states, "Marquette Long Lac Gold Mines Ltd. has been formed to acquire the southerly 14 claims of the 25 claims owned by Lafayette Long Lac Gold Mines Ltd. adjoining to the north of Little Long Lac."

The Marquette company's charter dated back to February, 1936, capitalized with 3,000,000 shares at $1.00 par value. Lafayette had sold its 14 claims for $3000. and 1,000,000 Marquette shares. H.J. Martin, President of Lafayette, and Peter Roche, a Director, became the President and Vice President respectively of Marquette. Marquette now held the claims on which Lafayette had performed work, TB 10943 (west zone) and 11321 (east zone).

Marquette embarked on a program to finance its operations. An ad in *TNM* on May 21 offered shares at 25 cents each, citing a report by W.W. Beaton which included the words, "The property is one of exceptional merit, none of the possibilities of which have been fully explored." His report alluded to that fabled mine only half a mile away, the Little Long Lac.

An ad by the same broker on June 4 stated that Marquette had awarded a diamond drilling contract. The next document on file that advances this saga is a typewritten report by Hans Lundberg dated August 11, 1936.

Much of Lundberg's report is highly technical. He described carrying out a geophysical survey on Claim No. TB 11222 (the "south claim"), the northeast corner of which touches the southwest corner of TB 11321 (east zone, now Kodiak's Goldstrike Zone). The survey ran along north-south lines cut 300 feet apart. The electrical survey used "the Ratiograph method", taking observations every 50 feet or less. The magnetic survey used "a sensitive variometer (Schmidt Balance, Askania type)". Lundberg produced two maps (not on file), with strong and medium intensities shown in bright red. He recommended two locations for drill holes.

Lundberg confined the survey to the vicinity of the iron formation in the northern part of TB 11222 (south claim), which also shows up in the southern part of TB 10943 (west zone). The northwest corner of TB 11222 touches the southeast corner of TB 10943.

He recommended exploration not in the iron formation itself, but in the area immediately north. "Close to the northern boundary of the claim, where a few outcrops of greenstone, sediments and diorite may be seen, the sedimentary contact has been geophysically well confirmed and in addition, indications of shearing and silicification have been obtained." Later information locates three holes in this area, M1, M2, and M3.

On November 21, mining engineer K.B. Heisey issued a report for Marquette. He stated that Holes M1 through M8 had been completed, for a total drill length of 2,021 ft. He also proposed another 5 holes. The results of the drilling program to date?: "No gold values of interest were encountered." He stated that previous drilling on the No. 11 vein (west zone) and in the shaft area (east zone) had been "productive of important results".

A letter dated September 30, 1936 from the field man-in-charge to the Marquette President and Directors, states in part: "Building materials are being freighted to the property, a carpenter and helper added to the payroll, and the camps are being put into shape before the freeze up." The references to building materials (which would include lumber) and freighting suggest that the property had road access. (Refer to the sidebar "The Lost Road" in an upcoming edition.)

Reports in *TNM* are scarce for 1936. An item on October 8 states, "The camps have been rehabilitated and preparations are being made for a continuance of operations throughout the winter months." An item on October 15 states that drilling had commenced. On November 19: "Scene of the present drilling is on the south claims [sic], across which have been determined two iron bands."

A map prepared by M.R. Brown circa 1945 shows Holes M4 through M8 in the northeast corner of TB 11222 (south claim).

There is a document on file that is the log of the core pulled from Holes M4 and M5. (Refer to the sidebar "Logs of Historical Holes" in an upcoming edition.)

A reliable source indicates that Marquette completed Holes M9 through M14. Brown's map does not identify Holes M9, M10, and M14. Holes M11, M12, and M13 are located on the No. 11 vein of TB 10943 (west zone). Heisey's report of November, 1936 recommended drilling "two [holes] in the shaft area and three in No. 11 vein area". That would account for Holes M9 and M10, likely drilled in the east zone. "These have been spotted on the ground," Heisey continued, "and M11 the first, is in progress." That accounts for M11, M12, and M13 on the No. 11 vein, leaving Hole M14 unaccounted for.

In January of 1937, two items in *TNM* confirm that Marquette was carrying out its drilling program. Thereafter, the information flow practically dries up.

View of Geraldton in 1937 looking north over Barton Bay. The engine of the community is the Little Long Lac mine, on the lakeshore. Errington Arena (now Errington Tim-br Mart) is under construction in the lower centre, beside the L.L.L. Townsite. A bridge spans the Bay.

Meanwhile, Geraldton was growing by leaps and bounds. A building boom in 1936 and '37 culminated in an application to become a municipality, and on September 1, 1937, the town was born. In 1938, both MacLeod-Cockshutt and Hardrock became producers.

The *Canadian Mines Handbook* has two scraps of information between 1937 and 1943. For Marquette: "Sept, 1937, oper on small scale only." For Lafayette: "Prop.: . . . Inactive in 1938."

From the *TNM*, an isolated item on December 7, 1939: "Lafayette Long Lac has been idle for several years, lacking operating capital." A brief item dated February 29, 1940 states that Marquette "expected that funds will be obtained for a further campaign of diamond drilling . . ." But nothing happened.

One overwhelming reality for the Little Long Lac gold camp was the advent of the Second World War in September of 1939. People rushed to serve, including miners. The first Geraldton casualty was a James Hendry, a MacLeod-Cockshutt employee. A news report at war's end listed 750 names, including some women, on the Geraldton District Honour Roll.

The preservation of history is sometimes a rocky process. The following is an extract from a letter from the President of Marquette to the government's district geologist in Port Arthur (Thunder Bay), dated November 11, 1946:

"Unfortunately a year or two before the war one of our Directors got all the reports of the drilling that was done under the supervision of Mr. Heisey, with his maps and records, and they have been lost. I was not responsible for these, they were given out by the Secretary and he never got them back, nor did the Director who lent them in the first place."

Not until 1944 did interest revive in the Marquette-Lafayette properties.

*ADDING UP THE NUMBERS

Although news reports, and even the *Canadian Mines Handbook* and some official government reports, cite the total drill length as 6,000 ft. before 1936, the numbers of the 20 holes add up to 4,975.4 ft., or 1,516.5 m. Heisey's report dated November 21, 1936 puts the total drill length at approximately 7,000 feet (4,975.4 + 2,021), or approximately 2,130 m. Holes M9 through M14 provided an additional drill length of 777.9 ft., bringing the total of Holes M1 through M14 to 2,800.9 ft. Therefore, the total drill length on the Marquette property in the 1930's was 7,774.3 ft., or approximately 2,370 m.

THE LOST ROAD

At some point, the Marquette Long Lac Gold Mines Ltd. property gained a road link to Geraldton and the CNR. A report in *The Northern Miner* dated March 28, 1935 stated, "[T]he camps have been reopened and a road cut in preparation for diamond drilling." We can be confident that the "road cut" was a winter road. Possibly it ran north from Barton Bay to the east zone (TB 11321). Some of it may have followed the hydro line corridor.

More likely it was the forerunner of an all-season road or trail heading generally west from Mine Road, starting in Johnsonville (Jonesville).

A year later, in a report dated May 4, 1936, W.W. Beaton stated, "The property is served with water transportation from Hard Rock, Ontario, on the Canadian National Railways, a distance of six miles, and by road from Geraldton, a distance of three miles."

So, a road existed in May of 1936. At that time of year, it would not have been a mere winter road, but an all-season road or trail, and to be a proper road or trail, it must have been built the previous year, in 1935, in the snow-free season.

What was the location of this road? There is no community memory of such a road. But there is a major clue about its location. On August 11, 1936, when Hans Lundberg reported on his geophysical survey of the Marquette property, he made this remark: "The property is located south west of Geraldton Station on the Canadian National Railroad and can be reached by 1½ miles of motor road from Geraldton Station and then by 1½ miles of trail to the west." One and a half miles is 2.4 km.

The one and only "motor road" from the station was, in 1936, Mine Road, now Main Street. An automobile odometer measures the distance between the tracks and Barton Bay as 2.5 km. The hydro line emerges from the bush at the Kilometre 2.2 mark. Between the line and Barton Bay, the terrain is unsuitable for a road or trail. In 1936, in the pre-odometer age, the "1½ miles of motor road" would have been an approximation.

On August 15, the writer undertook an investigation of the possible location of an early road on or near the hydro line corridor, beginning in Jonesville (just north of Barton Bay).

The corridor begins with a stretch of very swampy ground (long grass and alders) and a ditch, apparently excavated by the town many years ago for drainage purposes. Exercising some ingenuity, one can cross the ditch. There is no evidence of a road. The corridor runs straight west.

Almost a half kilometre from Main Street, a tote road emerges from the bush on the north side. Further along, a snowmobile trail crosses the corridor in a north-south direction. The corridor continues one kilometre from Main Street before it turns northwest. The ground is mainly grassy and soggy, with some brushy patches. There is no evidence of a road.

Proceeding northwest, one climbs a low sand hill, from which one can see another hill in the distance. Before reaching the second hill, one encounters a clearing on the west side, the area extensively stripped on Kodiak's Goldstrike Zone (Marquette's east zone). The distance traveled from Main Street is just under 1.5 km.

On August 15, 2008, Kodiak's Goldstrike Zone is easily visible from the hydro line corridor.

Retracing his steps, the writer searched the east and north sides of the corridor for signs of a tote road. One location near the Kodiak operation showed promise, but was heavily overgrown.

Back at the tote road identified earlier, the writer found a small clearing a few metres off the corridor. An overgrown tote road heads in a westerly direction. A road in good shape heads towards town. The clearing features a rusty engine block and an axle assembly, and the flattened ruins of frame structures.

The road has a good base. Just past the wreck of a sporty red car, the road is obliterated by thick brush. A footpath clings to the bank between the brush and the swamp on the north side. There is no bridge over the narrow excavated ditch that drains the swamp, but an ancient beaver dam can assist one in crossing.

An engine block rusts away in a clearing on the old Marquette trail.

A rough road begins (or continues) in a clearing, and as one proceeds east, it becomes Greer Avenue. Greer Avenue is 0.1 km north of the hydro line, i.e., 2.1 km from the tracks.

It is interesting that the tote road comes up on the latest GPS mapping software. Since no one and nothing has traveled this road for decades, its appearance on a modern map is very mysterious. The road on the map ranges westward and then stops abruptly in the middle of nowhere. Old National Topographical System maps, scale 1:50,000, show this same feature.

What was the purpose of this road? Herb Byerley, a longtime Geraldton resident, recalled that in the late 1940's, the road accessed a sawmill run by a local entrepreneur, Mel Fraser. He was manufacturing railway ties from jack pine.

In the report of March, 1935, the reference to a "road cut" may have been a trail from Greer Avenue to the hydro corridor. There would have been no ditch, and hence, no bridge necessary. Most of the winter road, then, would have run along the corridor.

In November of 1936, Lundberg referred to a "trail", not a motor road, running west from Mine Road. Even today miners refer to their "trails", which is a class below proper "roads", and a trail may accommodate anything from a hiker or an ATV to a 4WD vehicle.

Was the Greer Avenue Extension the lost road, the all-season route?

Only if it extended to the Marquette property. Modern technology has blessed us today with aerial photos of every square metre of the earth's surface. "Aerial" is not quite the correct term; they are satellite photos, taken from outer space. And anyone can view them online using a software program called Google Earth.

The writer used Google Earth, zoomed down to a height of a few

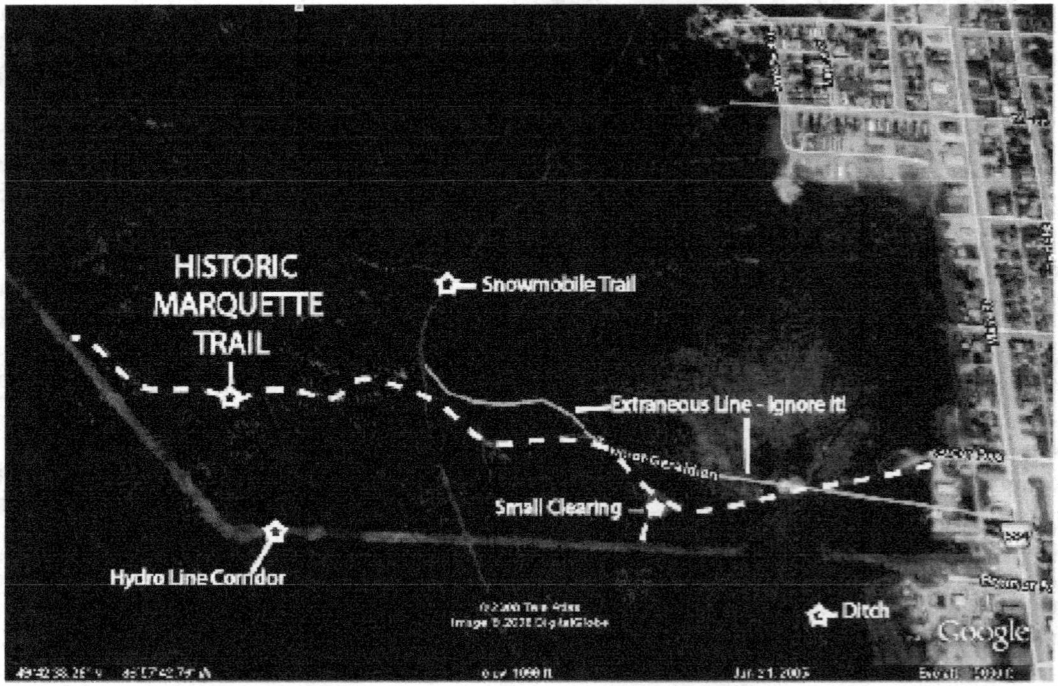

A Google Earth view from an altitude of 5000 ft. The trail accesses the old Marquette property

thousand feet above the terrain west of Greer Avenue. The trail can be traced easily to the clearing near the hydro corridor, and then it can be picked up as a brushy line heading westward, till it emerges again on the hydro corridor, just across from Kodiak's Goldstrike Zone.

The road is lost no more.

In the very early days in the Little Long Lac gold camp, the only way to transport freight overland was by sleighs in winter and wagons in summer. Teams of horses drew these conveyances to the Little Long Lac mine and to other prospects. Snowplows, if they operated at all, confined themselves to the area's major roads, such as they were, which always began or ended with Mine Road.

The Marquette trail would have been compatible with horse-drawn sleighs and wagons, and even tractor-drawn conveyances (At least one historical photo portrays this operation). The evidence of abandoned vehicles along that trail today probably dates back no earlier than the '40's.

As for the distance from Main Street to the old Marquette property, it is indeed less than a mile and a half. Even as an approximation it would be off by half a mile. But if the trail extended all the way to Claim TB 11222, where Lundberg was conducting his survey in 1936, then it was indeed a true mile and a half.

There may be an opportunity here for a local organization interested in opening and maintaining trails. Think about it. Think about the revival of the Old Marquette Trail.

View from the south side of the tracks at Geraldton Station, February 1934. Freight is transferred from boxcars to sleighs and shipped over winter roads to the mines. Photo by E.C. Everett.

WHEN DANGER LURKS

Deep trenches and open shafts are not the only dangers a tourist may encounter on an old mining property.

In late summer of 1940, a local prospector, Roy Barker, reported a hazard to the Ontario Provincial Police in Geraldton. On September 28, Constable E.M. Richardson enlisted help – he called in at the Little Long Lac mine to pick up A.S. Bayne, government mine inspector, who happened to be visiting at the time.

Richardson and Bayne proceeded to the Marquette property west of Geraldton's Main Street. They would have traveled the "lost road" from Jonesville. About "three miles" from town (Richardson wrote in his official report), they came upon "the old office building" (Bayne's words). The "three miles" might have been an over-estimation, for that would have brought them to the vicinity of the west zone on Claim 10943.

The reports by Bayne and Richardson raise some new mysteries. Where exactly was the office building? And why was the office building located so far from Marquette's main camp (Read further)?

Inside the structure they located the hazard. They had found three cases (150 pounds) of C.I.L. 50% Polar Forcite Gelatin dated April 16, 1935, and four sticks of 40% Polar Forcite dated November 8, 1935.

It was in November, 1935 that Lafayette, the owner of the property at the time, had made the decision to sink a shaft.

The bush, according to Bayne in his official report, was very dry, so burning the dynamite was out of the question. Bayne took two fresh sticks of 40% Forcite and two No. 8 detonators, and crimped two eight-foot safety fuse as primers. A single blast destroyed the hazard. "Detonation was complete," wrote Bayne, "and no trace of the explosives or containers were found to have been scattered by the blast."

Even though it would have been dangerous to move the old product, Bayne, according to Richardson, had moved the hazard to a clearing before destroying it.

Apparently Constable Richardson prowled the property looking for more hazards. A few days later, on October 7, he located another case of 50% Polar Forcite dynamite "about a mile farther in the bush from where the first was located in a shack with the roof caved in, the powder was very old and dangerous . . ."

On July 29, 2008, Steven Roach squats beside a trench on the west zone which the oldtime prospectors had blasted in the No. 6 vein.

Richardson raises more mysteries. Were "the old office building" and "a shack with the roof caved in" the same structure? Or was the last case of dynamite located in a shack with the roof caved in? And where does "a mile farther in the bush" bring you? It is known, by the way, that there had been some buildings near the No. 1 vein in the east zone of Claim 11321, and the chances are good some of these original buildings were constructed of logs.

Not having the means to explode the hazard, he returned on October 23. He brought with him "some fuse and caps that I located near the main Camp at the Lake" for the purpose of destroying them also.

Lafayette's main camp had been a cluster of buildings on the north shore of Barton Bay. It is possible that the constable had found the last case of old dynamite at this location too. The camp is only about a half mile south of the east zone on Claim 11321.

In concluding his report, Constable Richardson excused "George Papineau who is owner or in charge of the property" from any culpability. The property had no watchman, and the powder had been there a long time.

Part V – 1944-1951

While the Allies prepared to press forward to Victory in Europe, the *Canadian Mines Handbook* noted the following about the Marquette property: "New d d contract awarded May, 1944."

Both the Lafayette and the Marquette companies still had the same address on Bay Street, and the same President, H.J. Martin. The Marquette property comprised 11 claims, as before, and the Lafayette, immediately north of and contiguous to the Marquette, still had 14.

On October 19, 1944, a J.T. Burton, B.Sc., typed a progress report for Marquette, beginning, "The diamond drilling programme which commenced the middle of June on your property is now completing the 5th hole, intended to cross-section possibilities which might have been overlooked in previous campaigns."

The report referenced the No. 1 vein "in the area of the old shaft on Claim 11321 [east zone]" and the No. 11 vein on Claim 10943 (west zone). It outlined a new target: a shearing which trended approximately east-west about 200 feet north of the old shaft. An outcrop "some 500 feet west of the old shaft" exposed the shearing. The plan was to follow the shearing west towards the iron formation that Lundberg had identified near the north boundary of Claim 11222 (south zone). In that vicinity was a sedimentary contact, and north and west, the arkose in which the No. 11 vein occurred.

An undated map prepared by M.R. Brown showed Holes 44-2 through 44-6 following this plan. Another target was "a plug of intrusive diorite" 350 feet north of the shearing exposure. Hole 44-1, collared there, was bored generally south towards the shearing.

Burton reported that Holes 44-4 and 44-5 intersected diabase dykes, and that Hole 44-6 had been spotted near the iron formation so as to "cut the contact between the diorites and the arkose". He wrote, "...[T]he entire combination presents a picture of attractive possibilities. "

Burton also pointed out that "Little Long Lac Gold Mines find their values in a vein in the arkose . . ."

Burton went on to write that, in the event of disappointing results, "a few more short holes be drilled to trace vein #11 westward along its contact with the iron formation, as it is the only vein remaining on the property the merits of which have not been completely decided." Brown's map showed Holes 44-7 through 44-10 in that area. Another record[1] showed the last hole being completed on December 14, 1944.

Another geologist intimately familiar with the property was M.R. Brown. A letter dated October 23, 1946, signed by H.J. Martin, listed several communications from Brown starting with a progress report dated September 24, 1944. However, only two documents and the aforementioned map are on file. The letter also stated, "In 1944 the Marquette Company did quite a lot of drilling under the supervision of Mr. J.T. Burton and associated with him was Mr. M.R. Brown, M.E."

A letter from M.R. Brown, Mining Engineer, dated November 13, 1944, stated, "I must report that the assays from the intersection in hole 7 are very low. Naturally, I am bitterly disappointed. I have been sitting over in the assay office [in Geraldton] since 1 o'clock waiting for the returns. None will run more than .01 ozs. I was quite elated over the core when I wired on Saturday. It was unquestionably the nicest looking stuff we have had to date – in fact, about the nicest I ever saw . . . I want to get these assay results in today's mail."

The next communication on file from Brown is almost a year and a half later in time. Brown wrote a report for Lafayette Long Lac Gold Mines Ltd. dated March 9, 1946. We learn "the property is served by a good road from Geraldton, which runs approximately east-west across the entire length of the ground. This road was built in 1945 . . . " Elsewhere we learn the road was built by Lafayette in partnership with Marquette and Charles Long Lac Gold Mines Ltd.[2] The Charles property adjoined the Lafayette on the west. Today the road is known locally as the Blueberry Hill Road. It began in the vicinity of Pine Grove Cemetery at the end of Beamish Avenue West.

Brown also stated, "During the summer of 1945, the joint north-

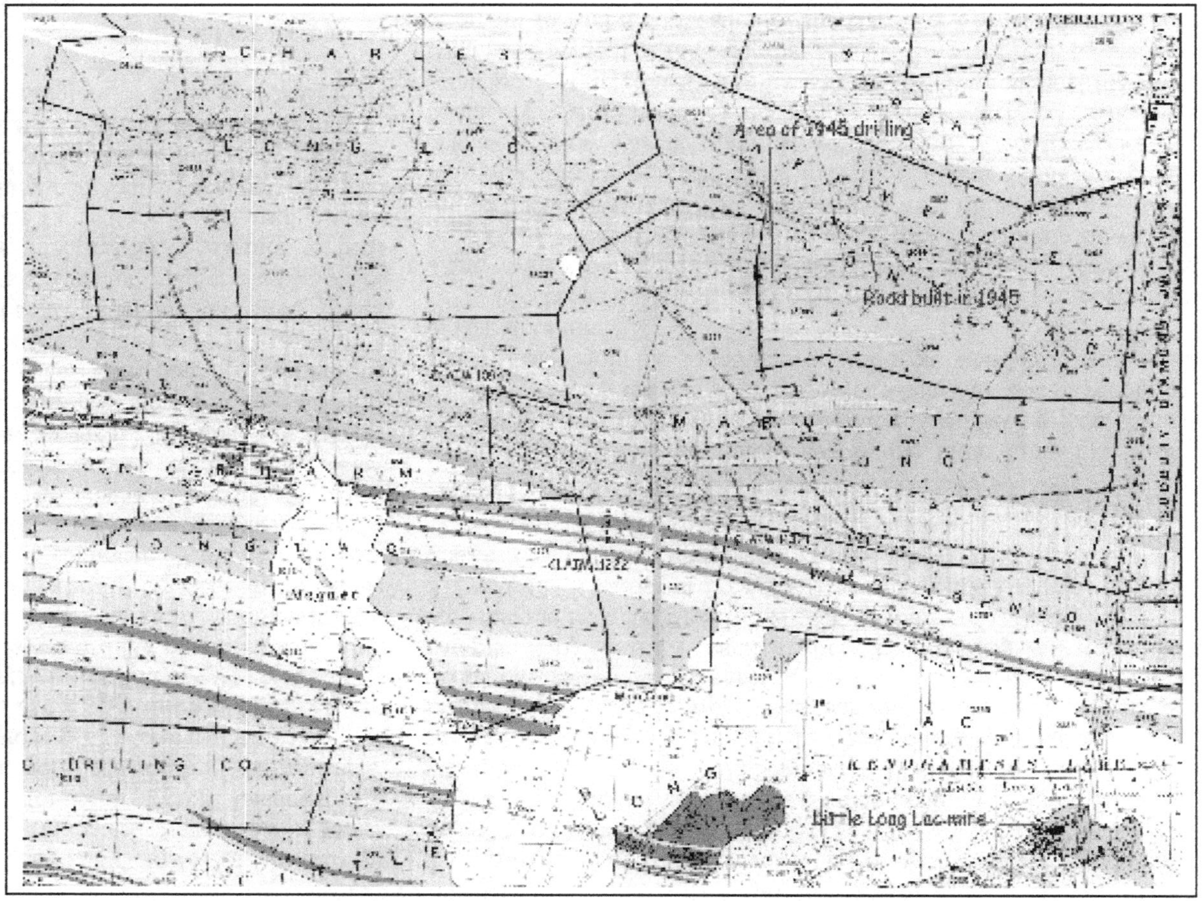

The scene of action in 1944-45, immediately west of Geraldton.

south boundary with the Marquette Long Lac Gold Mines Limited was cross-sectioned by diamond drilling." An Ontario Department of Mines map published in E.G. Pye's report[3] for 1951, showed unnamed drill holes east of the hydro line corridor, north and south of the new road. The target area was definitely not Marquette's historic target areas.

Another record disclosed that Holes 45-1 through 45-7, between August and October of 1945, had a total drill length of 4,914 feet.

Brown indicated that the Lafayette drilling program, undertaken jointly with Marquette (following the line of the north-south jog in their mutual boundary), was "the most northerly drilling done in the [Little Long Lac] camp proper to date." Brown continued, "Although this work did not locate any commercial gold values, it established what is considered an intriguing 'new' geological feature, . . . a major belt of sediments striking across the northern section of the [Layfayette] property. This parallels approximately the main Little Long Lac belt . . . Thus is indicated a major belt lying under an extensive swamp which is about fifteen miles long and a mile wide. This bounded on both the north and the south by high greenstone country."

Brown went on to describe the structure in more detail, concluding with "these sediments consisted almost entirely of interbedded greywackes and arkose, identical in appearance to beds found in underground workings of the Little Long Lac Mine." Brown pulled no punches: he stated that they had found no significant gold values, and that only "a great deal of work" would outline the structure properly. But the similarities to the Little Long Lac mine excited him.

Brown recommended another program of drilling, at least 5000 feet. We know today that nothing of the sort happened. Hole 45-7, completed on October 10, 1945, was the last significant exploration on both the Lafayette and Marquette properties.

An item in *The Northern Miner*, November 8, 1945, had this comment: "The drill has been dismantled and H.J. Martin, president, states that further drilling will have to await new financing."

In the spring of 2008, Kodiak Exploration Limited planted their flag on the old Lafayette-Marquette properties.

1 **NOTE:** Total drill length of 7 of 10 holes in 1944 = 3,915 feet, for an average of 560 feet/hole. An estimate for the missing data on Holes 44-6 through 44-8 = 3 x 560 = 1680 feet, for a 1944 total of 5,595 feet. Combine that with the total drill length of 7 holes in 1945 (4,914 feet), for a grand total of 10, 509 feet. E.G. Pye bluntly put the grand total at 10,600 feet (3,230.9 m).

2 **NOTE:** Charles Hutchison, of Charles Long Lac Mining Syndicate, staked the claims in 1935. In September, 1938, Charles Long Lac Gold Mines Limited was formed with 16 claims from the Syndicate. Some prospecting occurred, work was suspended during the war, and operations resumed on August 1, 1944. The E.G. Pye map showed camp buildings at the T-junction where the Blueberry Hill Road met a road that ran from the CNR tracks south to Magnet Creek.

3 Pye, E.G. "Geology of Errington Township, Little Long Lac Area", 60th A.R. of Ontario Department of Mines, Pt. VI, 1951.

MINING & POST-WAR GERALDTON

"The town of Geraldton, hub of the Little Long Lac mining district, is a thriving community once more after having gone through bleak days during the war when there was a heavy exodus of young men and women to join the armed services and more who left to work in the industrial centres."

So began an article in *The Northern Miner* in December, 1945. There were four operating mines: Little Long Lac, Hard Rock, MacLeod-Cockshutt, and Magnet. The mines were building crews up to full production strength. The vacant houses were filling up.

During the war, the forest industry had kept the local economy healthy. Now *The Northern Miner* predicted a mining exploration boom, for "there is a lot of worthwhile ground which has hardly been scratched." It noted that "iron formations stretch throughout the whole length and breadth of the camp but actually there has been no intensive effort yet to find a duplication of the Hard Rock-MacLeod north zone ore – and that is a prize worth winning." *The Miner* alluded to the promising new structure found on the Marquette-Lafayette properties.

In his 1951 report, however, E.G. Pye suggested why the boom did not occur. High mining costs persisted. The mines were finding a decrease in the grade of ore at depth. The price of gold, fixed at $35 an ounce, suffered when the Canadian dollar returned to par with the American dollar. Venture capital was still scarce.

Still, during the war, the local mining industry had soldiered on, handicapped by loss of manpower, rising costs, and scarce resources. By 1950, the Little Long Lac had produced 530,276 ounces of gold.

MacLeod-Cockshutt had produced 572,576 ounces, and Hard Rock, 259,708. Geraldton was still the hub of the mining district.

Geraldton circa 1950. An automobile enthusiast may be able to date the cars. Today, Stedman's occupies the Union Taxi location, and Rexall Pharmacy, the Thunder Bay Hotel lot.

POTENTIAL TOWNSITE

In his report of March 9, 1946, M.R. Brown alluded to good ground for a housing subdivision. First he described the Lafayette property as primarily "swamp and muskeg, sparsely treed by stunted spruce and tamarac." Then he continued, "A higher sand ridge strikes almost due west along the north portion of the property. It is along this ridge that the roadway [Blueberry Hill Road] has been built. This higher ground is heavily wooded with jackpine. A large proportion of this ground is included in the Town of Geraldton. Some of this ground may, in the future, be considered as choice building sites."

Longtime residents will recall that the ground at Hardrock Station on the CNR, east of Geraldton, was once touted as an ideal townsite.

– *Times Star*

TOUR OF KODIAK PROPERTIES

Part I – The Hercules and Brenbar

On Saturday morning, November 15, Kodiak officials conducted the writer on a tour of properties in the Jellicoe area. Jim Guillemette acted as videographer.

The temperature hovered just below zero Celsius, and a centimetre or two of snow blanketed the landscape. At the Kodiak office at Bush Lake Resort on the 801 Hwy., Robert "Bob" Hawkins presented an overview of operations. Hawkins, Chief Geologist, pointed out that his company has holdings of over 1,400 square kilometers in the Beardmore-Geraldton gold belt.

Hawkins demonstrated the 3-D modeling software using drill hole data from Kodiak's flagship Hercules Project, which contained the promising values of the Golden Mile Zone. Kodiak was still in the exploration phase on all its properties, scattered over a length of 150 km. "Typically," said Hawkins, "an underground mine has to have 0.3 ounces average grade [10 grams/tonne]." Later on the tour he commented, "Our target is [at least] a million ounces, something to justify a mine and a mill."

Hawkins and Erik Haroldson, project geologist for the Hercules, led the party to the core "shack", a one-and-a-half-story building where several workers logged and cut diamond drill core. At one point, a helicopter flew in with a sling full of core from the Maki Midas Property. Because the moisture-laden air was freezing on the chopper's windows and decreasing visibility, a helicopter tour of Kodiak's properties had to be postponed.

Jumping into a pickup truck about just before 10:00 o'clock, Hawkins, Haroldson, Guillemette, and the writer left for the Hercules Property, utilizing the 801 Hwy. and the River Road. At the Kilometre 29.5 point on the Kinghorn Road, the truck turned left (west) on an old one-lane bush road, which Haroldson called the Back Road. The truck bumped along for over 4 kilometres, stopping in a jack pine plantation, just south of Golden Pond, a previously unnamed lake. In 2007, the Golden Mile had been stripped up to that lake.

The Golden Mile trends northwest for about 3 kilometres from the Kinghorn Road. This year Kodiak had discovered a parallel gold zone, about 500 m southwest of the Golden Mile, calling it the Lucky Strike. Now a rig was positioned here to bore the longest hole to date (HR08-297), at a bearing of 33 degrees (northeast), to intersect both gold zones.

Erik Haroldson (left) and Bob Hawkins inspect a rig on the Hercules Property that is positioned to drill Kodiak's longest hole, HR08-297.

At the time of the tour, the rig had shut down while the drillers fussed with the hoses for water supply, possibly because the onset of cold weather had caused some problems. Hawkins explained, "This is [still] exploration drilling; we're not defining a resource yet."

At the height of the past season, Kodiak had had 6 rigs operating on the Hercules Project; now there were two. Over the holiday season, the rigs would be withdrawn, and a few key people would analyze a backlog of Information from many sources – channel samples, drill core, geophysical data. "It's a matter of sitting down," he said, "and taking all this data and deciding where to go next."

On December 3, Kodiak released a table of information on all the holes which it had drilled on the Hercules in 2008, and for which assays had been returned. This year Kodiak had drilled over 62 km total length on the Hercules alone, the deepest down hole depth being in the range of 600 m. Hole HR08-297, however, which the touring party had visited, had reached a down hole depth (as opposed to a vertical depth) of over 1000 m, but assays were still pending. A visual inspection of the core at a depth of 1,027 m showed great promise.

That deep test hole had been bored in Area 51 of the central portion of the Golden Mile. Higher in the same mineralized high grade shoot, another hole, HR08-192, had intersected 75.3 g/t (2.20 ounces per tonne) over 0.4 m at a down hole depth of 231 m. The plan, Hawkins explained to the party, was to "wedge off" the long test hole to find additional deep intercepts.

The news release also stated that Kodiak had been drilling this season beyond Golden Pond, to the northwest, and had conducted surface exploration for 5 kilometres, all the way to the boundary with Sage Gold Inc. Sage recently announced good values on its Golden Extension showing, on strike with the Golden Mile. Kodiak has now reported evidence of an additional strike length of 5,000 m for the Golden Mile, and plans to conduct a drilling program there this winter.

Next, the party inspected the Marino Zone, excavated in early 2007. Except for the Derek Lee Zone (which was scheduled to be done later), every zone between the Penelton Gold and the Marino Zones had been reclaimed, meaning that the excavated material has been returned to the trenches so that nature could do its work of vegetating the disturbed areas, a total of 45,600 m^2.

The party returned to the 801 Hwy. At a point exactly 11 km from Hwy. 11, the truck turned west on an old one-lane road that accessed the properties of the old Sturgeon River and Brenbar mines. Back in the thirties, these properties had been hot prospects, and money and shafts had been sunk into them. Kodiak has an option to purchase the Brenbar, originally known as the Casey Contact Gold Mines prospect, and subsequently as the Brengold mine and since 1945 as the Brenbar mine.

The road passed a small trenched site and almost immediately ended in a small clearing with a pole tent frame and a concrete-capped shaft whose dimensions measured approximately 2 by 3 m. Hawkins led the party through the brush about 30 m north to an open pit filled with water, an uncapped shaft. Scattered through the bush were ruins of buildings, he said.

Beyond the open shaft on the Brenbar Property, the stripped shear zone climbs the hill.

Beyond the pit a stripped area climbed the hillside. Beginning about 3 months previously, Kodiak had followed the quartz vein through which the historical open pit had been sunk, and stripped a broad shear zone about 400 m in total length.

"We got the results two days ago," said Hawkins, referring to channel samples. The results had not been officially released, but he did say that 2 to 3 g/t was a good indicator. He said that he planned to trench more in the spring, and to spot drill in order to intersect the vein. "That's what Brenbar is," he said, "a big wide shear zone like this."

What prompted Kodiak to explore Brenbar? "We knew there was an old mine here, we knew there was a quartz vein with gold." The quartz vein is geologically younger than the shear, so, in effect, the vein has concentrated the gold it encountered in the shear. "It gives me a target," he continued, "a place to look to see if it is better than that. So I'll be drilling below that, and alongside of it."

The historical production of the Brenbar is unclear; one figure is 63 ounces; another is possibly 155 ounces. The Sturgeon River right next door produced 73,438 ounces.

The party returned to Bush Lake Resort to have lunch, and to determine if the weather would permit a helicopter tour of Kodiak's other properties.

Bush Lake Resort. The story-and-a-half building on the right is Kodiak's core shack. The bungalow at centre, with the deck, is Kodiak's office.

Part II – From Maki Midas to Solomon's Pillars

About 1:30 p.m. on November 15, the weather having improved, pilot, Francois "Frank" Brisebois removed the sleeves protecting the blades of the Bell 206L helicopter from the accumulation of moisture. Shortly thereafter, the chopper lifted off the ground in order to shuttle the touring party – Bob Hawkins and Erik Haroldson of Kodiak, and Jim Guillemette and the writer – to the Maki Midas property.

Brisebois was counting on the weather window remaining open for the round trip. Maki Midas was just a few minutes' flying time in a southerly direction from Bush Lake Resort, south of Hwy. 11 and the Blackwater River.

Aerial view looking generally east up the trench. Objects in view are the backhoe (centre, lowest object), two tent shelters, and the drill rig.

From the air one could see a trench in the heavily wooded area running generally east-west for a kilometre. A drill rig was positioned south of the trench. A little further west, a backhoe was clearing an area around the collar of the next hole to be drilled.

Brisebois landed the chopper expertly on the north side of the trench just metres from the excavation. Immediately Hawkins noted that the rig was not running. It had encountered some problem on Hole MK08-03 on this second day of operation. Last winter, Kodiak had drilled its first two holes on the property.

Standing on the bedrock in the shallow trench, Hawkins explained that the targets were banded iron formations in a series of parallel zones aligned east-west, and stretching south for over 1.5 km. The host rock was basalt or greenstone, its surface contoured by continental glaciers, in the same structure in which the old Northern Empire mine, further to the west at Beardmore, had found its ore. Roxmark Mines Ltd. was in process of exploring that property.

"The banded iron formation is actually the host for structures that hold the gold," said Hawkins. On a geological map, one can follow the structure all the way east to Geraldton, where it passes south of the old MacLeod-Cockshutt mine. Hawkins explained that Kodiak's Sturgeon Bridge and Solomon's Pillars properties were on parallels structure to the north, and that all of the structures were within the same structural belt on which the MacLeod-Cockshutt and Hardrock mines had found their ore. Prospectors have been poking around the Maki Midas property since the 1930's. There have been no big finds. "You see," said Hawkins, "nothing's gone on over here because that river has been a barrier to exploration." He gestured north to the nearby Blackwater River, hidden from view. In the fall of 2008, the Ministry of Natural Resources had removed a key access to the Maki Midas when it demolished an ancient, unauthorized bridge over the river. Having to fly in all equipment and supplies – at least until an ice bridge formed – had doubled the cost of drilling, according to Hawkins.

Bob Hawkins examines a mineralized specimen from a boudin (a swelling in the BIF).

Hawkins pointed to the banded iron formation snaking along the 20-metre-wide trench, from 1 to 3 metres wide in pinch-and-swell fashion. "That's a minable width if we can get gold values consistently," he said. Earlier Kodiak had finished its surface exploration on this trench – a geophysical survey, geologic mapping, and surface sampling. On Hole MK08-03, the first hole this fall, the drill had reached 80 to 90 metres downhole.

Expanding on the possibilities of the property, Hawkins gestured towards the west. He commented that a prospector 5 kilometres away had uncovered an identical formation and found good values. Gesturing toward the south, Hawkins said, "That's why I like this structure – it has multiple targets." He said Kodiak could sink a shaft in the middle and explore in all directions.

Hawkins got down on his knees to examine the rock. He picked up a chunk and commented, "It's intensely silicified banded iron formation [BIF] due to the mineralized shear zone that is hosted in the BIF."

Erik Haroldson (left), Bob Hawkins, and Jim Guillemette check out the banded iron formation (BIF) on the Maki Midas.

When the party returned to the chopper, the pilot, Brisebois, announced that the weather had improved enough that the flying tour could continue. First, to give the passengers a view, he flew the length of the trench from west to east. Continuing east, the chopper flew over Kodiak's Blackwater property, a 500-metre trench just before Jellicoe. There had been no drilling program.

The next destination was the Sturgeon Bridge property, named for the CNR bridge that spans the Namewaminikan (a.k.a. Sturgeon) River where it exits Partridge Lake. One could see a bush road on the north side of the river, starting in the vicinity of the Kinghorn Road bridge, running east toward the property, part of which encroached on a clearcut. Hawkins said the structure was a shear zone, similar to the Brenbar and Brookbank properties. Results from the 9 drill holes were pending.

Kodiak's news release on December 3rd reported that the Sturgeon Bridge-Caviar zone "discovered multiple areas of mineralization, including high grade gold mineralization: 0.6 metres grading 66.7 gpt (1.95 opt) in drill hole SB08-13".

The chopper changed direction, flying generally northwest. On the north side of Atigogama Lake, the old Orphan mine showed evidence of clearing and stripping. Mantis Mineral Corp., said Hawkins, was exploring host rock of the same age and composition as the Elmhirst Lake Intrusion on Kodiak's Golden Mile, and similar vein structures.

Further west at Kaby Lake, adjoining the Mantis property, Kodiak's Gold Rush Zone revealed about 300 metres of trenching in the same host rock and with a northwest trend similar to the Golden Mile, 13 kilometres to the west. Results from 6 holes were pending.

On the return trip to Bush Lake, the party flew over the Solomon's Pillars property, accessed by the road that passes by the TransCanada Pipelines pumping station just west of Jellicoe. Kodiak had stripped about 500 metres to examine altered sedimentary rock. However, 19 holes had returned disappointing results, and Hawkins said they would be dropping their option and reclaiming the property (returning it to a state of nature).

The chopper touched down at Bush Lake at 3:15 p.m.

Kodiak has announced that it plans further exploration on the properties seen on this tour, with the exception of Solomon's Pillars. Other active properties not visited recently are West Geraldton, East Leitch, Coté Archie, and GMO.

- Times Star

The capped shaft at the Brenbar Property

The drill rig on the Maki Midas

The Sturgeon Bridge Property

KODIAK

EXPLORATION LIMITED KXL: TSX.V

Kodiak Exploration Ltd. – 1205-700 Pender St.
Vancouver, BC V6C 1G8 – info@kodiakexp.com

Kodiak Exploration Limited (KSL : TSX-V) is a North American-based junior mineral exploration company that acquires and explores undervalued mineral properties with world-class potential. Kodiak's primary property, the Beardmore-Geraldton gold camp, is located 210 kilometres northeast of Thunder Bay, Ontario, Canada. This region consists of over 2100 km^2 of 100% controlled claims and numerous optioned properties on high-priority targets. Within this past-producing gold belt, Kodiak has made a significant gold discovery on its flagship Hercules property. The Golden Mile and multiple parallel gold-bearing vein systems hold tremendous resource potential.

In addition to the resource potential of its Beardmore-Geraldton gold properties, Kodiak's energy division is actively pursuing uranium exploration on two key properties in Quebec and Saskatchewan. Kodiak's Otish mountains property hosts a newly discovered uranium-bearing structure. The 2009 exploration program for this project will determine the extent and grade of this structure. In Saskatchewan, Kodiak is drilling high-quality and compelling uranium targets on its West Millennium property, a property surrounded by the tremendously rich uranium deposits of the southeastern part of the Athabasca Basin.

Kodiak is a well-funded and institutionally supported company with a strong management and professional team that conducts systematic district-scale exploration programs. With a proven track record of discovery, Kodiak is prepared to move forward on multiple exploration fronts, advancing its projects towards feasibility, making new discoveries, thereby growing the company and increasing shareholder value.

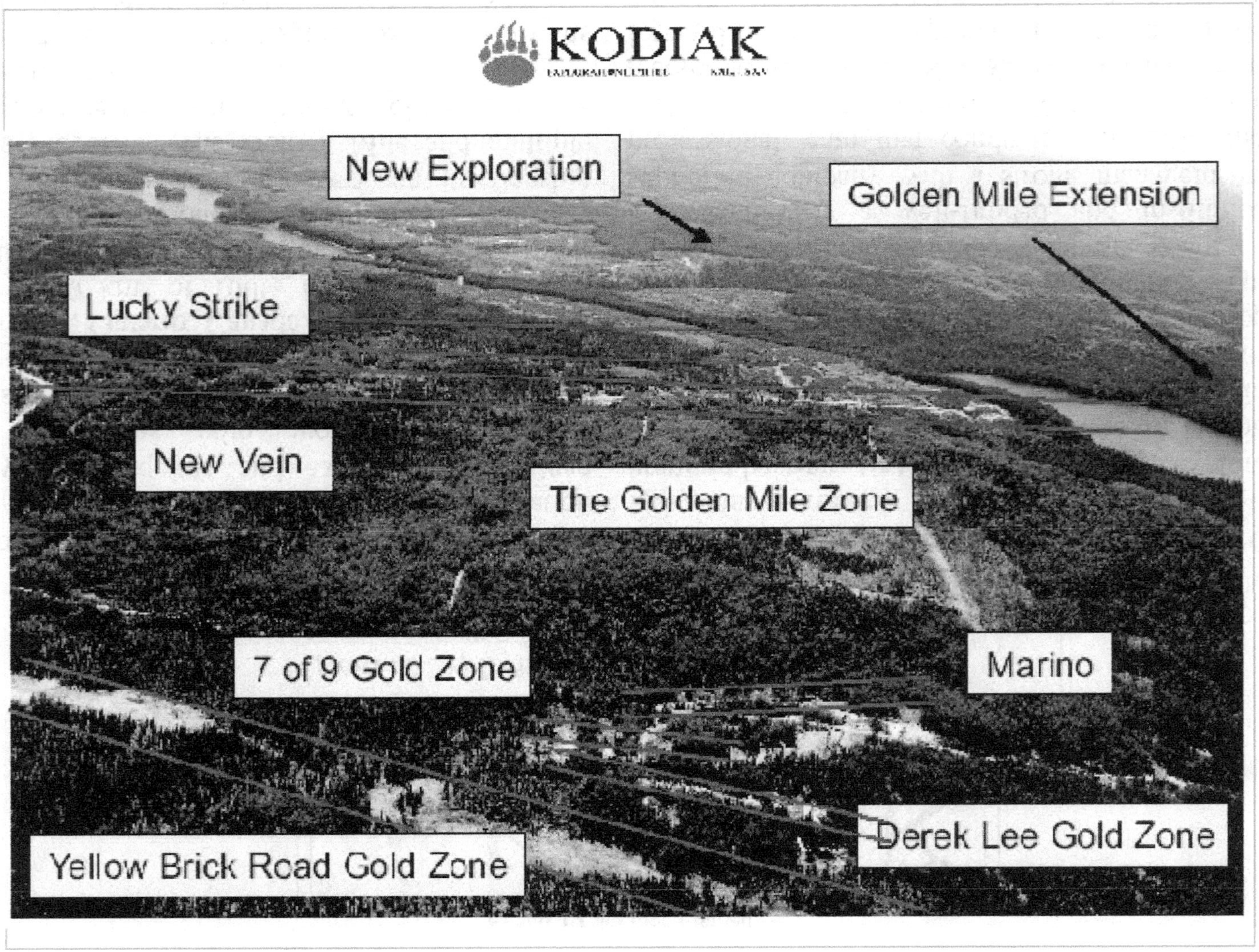
KODIAK
New Exploration
Golden Mile Extension
Lucky Strike
New Vein
The Golden Mile Zone
7 of 9 Gold Zone
Marino
Derek Lee Gold Zone
Yellow Brick Road Gold Zone

KODIAK

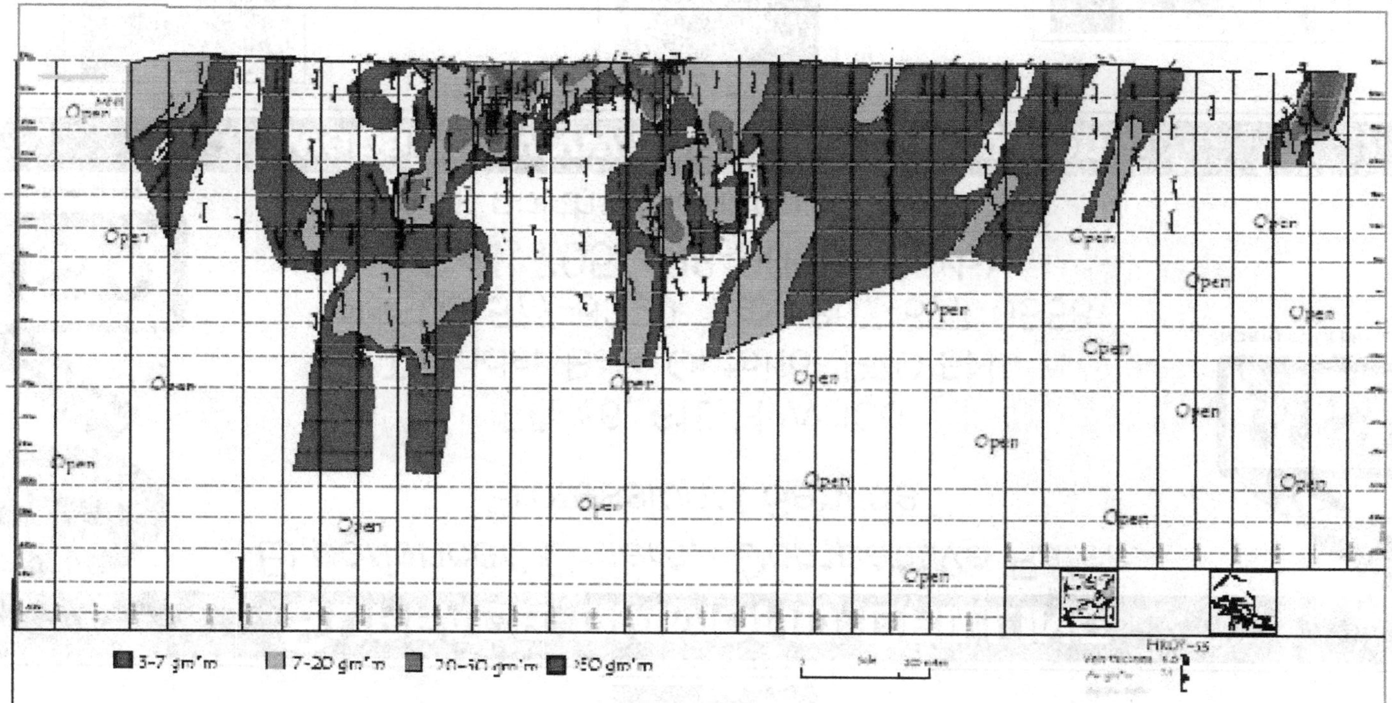

Long Section Golden Mile

Hercules Project

Au Grade Thickness - Golden Mile

Kodiak Exploration Ltd.
1205-700 W Pender St.
Vancouver BC V6C 1G8

SURFACE DIAMOND DRILLING

We've Moved!

Experienced Drillers, Competitive Rates, Professional Service

RR #6, 615 Hwy 130
Thunder Bay, Ontario P7C 5N5
Tel: 807-577-5399 Fax: 807-577-0558
Cell: 705-498-9056 (Barb)
email: northstardrilling@tbaytel.net

NSD NORTH STAR DRILLING LTD.

website: www.northstardrilling.com

Chapter 2 :
Year 2008

Activities of Municipality of Greenstone

Muskeg Tours 2009

View from the Palisades of the Pijitawabik

KODIAK OPTIONS FROM GREENSTONE

The Municipality of Greenstone is now a partner in a mining venture. Kodiak Exploration Ltd. recently optioned a block of claims from the Municipality just west of Geraldton's residential area.

On June 9, the Municipal Council passed By-Law 08-21 to authorize an option agreement with Kodiak for mining properties. As Kodiak explains in Schedule 'A', "Kodiak holds title to a large block of mineral claims contiguous to the northern and western boundaries of the patented mineral claims held by you just outside Geraldton".

"Accordingly," the Schedule goes on to say, "Kodiak is pleased to offer to acquire from you an option (the "Option") to acquire from the Municipality of Greenstone (the "Municipality") 100% of the Municipality's right, title and interest in and to the mineral rights only of the following patented claims situate in Errington and Ashmore Townships, Thunder Bay Mining Division (the "Property"):"

A list of 30 mining claims follows, among which is TB 10943 and TB 11321. On these two claims Kodiak is currently conducting surface exploration in the "Blueberry Hill" area, as the area is locally called.

Greenstone retains the surface rights on the patented claims, and if the option is exercised and fulfilled over five years, until October 31, 2012, and a mine is established, the Municipality will have a 2% net smelter returns royalty.

In addition, Kodiak agrees to pay the Municipality an undisclosed amount of money each year that it exercises the option. The amount is not yet public information, but the street buzz is that hundreds of thousands of dollars are involved.

The first payment, described as a "goodwill payment", has already been made. Annual payments are due on October 31, starting in 2009. "The dates proposed for future payments will allow Kodiak the time to conduct that initial work and to conduct its future work programs during summer months."

Other than the initial goodwill payment, all payments will be made at Kodiak's sole discretion. "Kodiak may terminate this Agreement on prior written notice to the Municipality at any time" hereafter. If Kodiak fails to make a payment, or to honour the royalty undertaking, the Agreement is deemed to be terminated, and Kodiak will forfeit whatever it has paid to that point.

On claim TB 10943, Eric Deroy of Geraldton uses a channel-cutting saw to extract a sample that will be sent out for assay.

Speaking to the Times Star, Roy Sinclair, CAO for the Municipality, explained that in 1984, the then Town of Geraldton annexed contiguous areas in portions of the Townships of Errington, Ashmore, Fulford, and McQueston. At that time it acquired jurisdiction (but not ownership) over several patented mining claims.

Subsequent to the annexation, the estate of the late Roy Barker approached the Town with an offer of sale of patented mining claims on which the estate was paying annual taxes to the Province. The Town bought the claims, and in the documents, it was pleasantly surprised to learn it had acquired the mineral rights along with the surface rights.

Today, of course, the former Town of Geraldton is now a Ward in the Municipality of Greenstone. Earlier this year, Kodiak Exploration Ltd. approached the Municipality with an offer to acquire the mineral rights to a block of claims in the bush west of Geraldton.

Sinclair scrambled to learn what he could about option agreements, and the Municipality hired the services of a Toronto lawyer who specialized in that area. As required by law, the Municipality advertised its offer of mineral rights, and only Kodiak responded. Kodiak's lawyers and Greenstone's lawyers hammered out an option agreement.

With permission, Kodiak began work on the block of claims in May, before the formal authorization of the agreement by Council on June 9.

Other articles deal with the exploration activities on these claims, and delve into the history of the property. *- Times Star*

GREENSTONE

A Rich Mining History - A Golden Mining Future

(Beardmore, Caramat, Geraldton, Jellicoe, Longlac, MacDiarmid and Nakina)

THE CORPORATION OF THE MUNICIPALITY OF GREENSTONE

HERITAGE INTERPRETIVE CENTRE – ARCHITECTURAL CONCEPT OF DRILL RIG

The Geraldton Heritage Interpretive Centre sits astride reclaimed mining tailings and is an architectural design concept of a drill rig. Drill rigs currently are operating across the highway from the Centre.

This was one component of a multi-million dollar mine tailings reclamation project by Barrick Gold in concert with the former Town of Geraldton. Another component by Geraldton was the creation of a 9-hole golf course constructed totally on mine tailings.

Refurbishing of the last standing headframe was another important project component to ensure the history of mining in the Beardmore – Geraldton gold camp.

Geraldton Heritage Interpretive Centre

Interior of Geraldton Heritage Interpretive Centre

PARTNERING IN MINERAL DEVELOPMENT

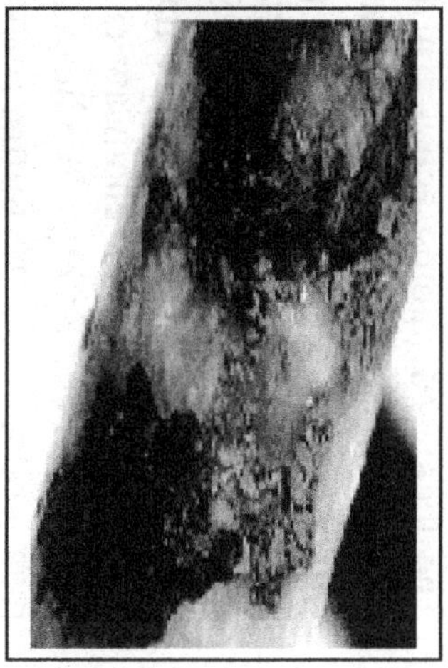

Greenstone will continue to be supportive of mining initiatives just as the former municipalities of Beardmore, Geraldton, Longlac and Nakina have always been in the past.

In 2008 Greenstone advertised for mining companies to submit proposals for Option Agreements for mineral rights that the Municipality owns just north of Barton Bay.

Greenstone subsequently entered into an Option Agreement with Kodiak Exploration Ltd. to verify and expand significant drill hole results from the 1930's. Expenditures to date on these properties are in the excess of $1.1 million.

Further exploration is planned for 2009.

Geraldton Community Forest (GCF) was established in 1993 to further the concept of sustainable forest management and to seek out employment opportunities supporting local economies. GCF employs and trains over 200 skilled workers from the area including several First Nations communities during peak times of the year. Heavily involved in the Provincial Forest Fire Program for the last decade, the company has over the last few years diversified and delved into new areas of business including GIS (Geographic Information Systems) and IT (Information Technology).

Geoma, the GIS division of GCF has established itself as a major driving force in the GIS industry providing services throughout Ontario. The information technology department in conjunction with Geoma has implemented cutting edge technologies and a variety of innovative products enhancing the way data is archived, displayed and distributed to the world. We recently completed a mineral development project in partnership with the Municipality of Greenstone and the Greenstone Economic Development Corporation.

Geoma was responsible for developing an innovative mineral exploration model using information and communications technologies (ICT) and GIS to study an area of high base-metal and precious metal mineral potential within Summers and Eva Townships in the Beardmore Ward of the Municipality of Greenstone. The initiative resulted in providing easy access to information in an attempt to improve competitiveness in attracting mining sector exploration.

The project scope was developed taking into account that geologists are now basing mineral exploration and assessment activities on the scientific foundations of mineral deposit models. Advances in Science and Technology, such as GIS and three dimensional (3D) modeling, reflect fundamental changes in how users define and use computer systems. The best mineral potential GIS evaluation models combine both data-driven and knowledge-driven geologic information. For more information on the project or to view the final product visit www.greenstone.ca/gmdi

The increased mining activity in the area has provided new avenues of work for GCF. A number of exploration companies over the past few years have hired us to provide training in first aid and CPR based on the St. Johns Ambulance certification, forest fire fighter training, professional chainsaw certification, WHMIS, TDG, and Health and Safety. We have also provided GIS services, map production, large format scanning, GPS data collection.

GCF is a partner in the Greenstone Community Portal and manages the Greenstone Datacentre on behalf of the Municipality. The focus is on empowerment of businesses and organizations through collaborative development. Services offered include a full range of software and equipment hosting and professional information technology consulting. Services are free where supporting not-for-profit activities benefitting the Greenstone area, and of minimal expense otherwise.

GCF at Sudbury Mining Conference

Michael Gravelle, MPP

Thunder Bay—Superior North

"Working together to celebrate the Mining Industry and the continual growth of the Greenstone Area!"

101-215 Van Norman St., Thunder Bay, ON P7A 4B6
P: 1-888-516-5555 W: www.michaelgravelle.ca

Chapter 3 : Year 2008

SAGE GOLD INC.
Precious and Base Metals in the Americas

Activities of Sage Gold Inc.

Muskeg Tours 2009

Approaching the Golden Extension

SAGE'S PROPERTY PORTFOLIO
Part I – The Golden Extension

The forest industry does help mining exploration. Sage Gold Inc. has found what it believes will become a major gold discovery in the middle of a clearcut.

Aerial view of Sage's Golden Extension. The line of strike (right to left) points to Kodiak's Golden Mile Zone about 5 kilometres away. MNDM Photo.

When Sage's prospectors made the discovery, they did not have to fight their way through swamps and bush – at least in that location. And they found a logging road in place.

On September 22 and 23, the writer joined a field tour of some of Sage's properties in the Beardmore-Geraldton mining camp. The party included Sage officials, a group of investors from Bay Street, and two geologists from the Ministry of Northern Development and Mines (MNDM). The four investors – perhaps mining investment advisors is a more accurate term – had flown into Thunder Bay Airport that morning, and a chartered bus had whisked them to a tourist resort on the 801 Road between Beardmore and Jellicoe.

The first day began with a lunch in the dining hall of Pasha Lake Cabins and a presentation by two of Sage's top executives: Nigel Lees, President and CEO; and Bill Love, Vice President/Business Development.

The field party jumped into 4-WD pickups, two of them driven by local prospectors, Myron Nelson and Lyle Holt, whom Lees also described as "partners" in Sage because they have vested interests.

One truck scooted over to nearby Bush Lake Resort, where Sage locates its field office, so that a small group could catch a helicopter ride to the Hercules Property of Kodiak Exploration Ltd. Four trucks headed north on the 801, cut across to the Kinghorn Road by the River Road, and entered the Kodiak property.

The trail into the Golden Mile Zone had been improved since last year, and a rough extension had been built to the north end of the area stripped of overburden, running on the west side. After a long wait, two helicopters touched down in the clearing. Out of the Kodiak-sponsored chopper stepped one investor and four Kodiak officials. The Kodiak team planned to shuttle parties of investors all week to its star property.

As pre-arranged, the Kodiak team, led by Bill Chornaby, President and CEO, and Brian Maher, Vice President/Exploration, was going to make a presentation to the Sage/Kodiak parties. Everyone migrated toward the "jewelry box" of the Golden Mile and descended into the shallow trench. A little way to the south a North Star Drilling rig groaned away.

It was a day that held a lot of promise, in more ways than one. On this warm autumn day, the sunlight bathed the wide quartz vein that snakes through a shear zone in the host rock, the Elmhirst Lake Intrusion. Brian Maher got down to business, spreading generously sized charts on the gold-bearing vein and speaking to the graphics contained therein.

Since late summer of 2006, Kodiak had been making spectacular gold discoveries on its Hercules Property. The Golden Mile Zone had been stripped almost continuously for about 3 kilometres, trending northwest. This season diamond drills had been sinking deep holes in a continuing effort to define an ore body, a substantial mineral resource that would justify a mine.

In the first two seasons, Kodiak stripped a parallel line of zones to the east, running from the Wilkinson Lake Zone in the southeast to the Marino Zone in the northwest, This season it has been stripping and drilling a parallel zone to the west. This past summer, said Chornaby (President), Kodiak had 135 people employed in the field.

Maher (VP) stated that Kodiak was sending about a thousand core samples a day to the lab for "fire assay". Any result yielding 3 grams per metric tonne (g/t) or better had been sent back for a more rigorous "metallic assay". Drills had intersected values as deep as 575 metres. Major mining companies have expressed interest, but, Maher said, negotiations were confidential. Holding up his hands, Maher remarked: "We have almost two handfuls [of documents] signed."

In the course of the presentation, Chornaby said, "Our neighbours [Sage Gold] are on strike with a significant discovery." He was referring to the Golden Extension Showing about 5 kilometres across country.

Later, Gerry White, district geologist for MNDM, communicated to the writer: "If Sage's Golden Extension occurrence proves to be the northwestern extension of Kodiak's Hercules vein system, a much large exploration target emerges from this area."

Squatting in the "jewelry box" of Kodiak's Golden Mile, Brian Maher (VP) addresses the field parties. Sage's President, Nigel Lees, looks on (upper left).

After this splendid example of cross-promotion by the two junior companies, the two parties dispersed. Most of the Sage field party embarked in the four pickups. The writer and two others climbed into the Sage-sponsored Bell 206 Long Ranger and flew northwest along the strike of the Golden Mile.

To the west, one could observe the newest Kodiak zone, the Lucky Strike, also trending northwest. Ahead, just north of the unnamed lake that terminates the trenching on the Golden Mile, drills were churning away in the bush, accessed by a rough trail that permits only ATV travel. Farther ahead, a clearcut area emerged.

Up to a week ago, the Sage crew had been stripping the vein by hand. With a bridge newly installed over Fairview Creek, operations were continuing with a backhoe. As the helicopter landed, other members of the field party were making their way by road. A logging road runs west for almost 11 kilometres to the showing, starting just past the Kilometre 43 marker on the Kinghorn Road.

The backhoe operator, Nolan Cox, greeted the visitors, along with son Ted Cox, Sage's field geotechnician, and son Danny Cox, contracted by Sage to provide a crew of geotechs.

Trenching had exposed a broad quartz vein in a shear of the host rock, the Elmhirst Lake Intrusion. Bill Love (Sage VP) explained, "The strike of this is very close to that of [Kodiak's] Golden Mile."

The Golden Extension Showing occurs about 1.6 kilometres from the boundary with Kodiak. The vein has been traced over a length of 200 metres, most of it in the last week. Ted Cox described the gold-bearing vein as a tight pinch and swell structure, the widest point being 1.3 metres, with a vertical dip. Gazing to the southeast, Cox said, "We can hear Kodiak's drills every day."

Cox wowed the visitors with a demonstration of panning. Scooping some soil and dipping the pan into water, he gently rocked and swirled the contents until a glistening tail of gold appeared. Said one of the investors, "It's amazing! That gold just came out of the muck [near the vein]." Donna McPherson, a stock broker, represented Octagon Capital Corporation, based in Toronto.

Walking with some visitors northeast of the main trench, geotechnician Cox gestured towards a new water-filled trench. He explained that in developing a "waterhole" to supply the hose that washes down the bedrock, the excavator exposed the vein extension with startling samples of visible gold (VG). The strategic plan of operations, said Cox, includes some cross-trenching to determine if the showing has some parallel zones alongside.

Ted Cox demonstrates with a miner's pan how to extract gold from the muck.

The Cox family has a financial stake in the Jacobus Property, which is an extensive area northwest of Kodiak's Hercules Property. The Golden Extension is located in the Jacobus East Area. Danny Cox, a prospector and an exploration contractor, remarked that their financial interest gave them an extra incentive to do their jobs well.

Back at the main trench, the rest of the field party, along with Nigel Lees (President), arrived in three trucks. One truck had foundered in a muddy section of trail near the Kodiak drill rig, and the drilling crew had helped extricate the vehicle. Another problem had developed, and they had abandoned the truck temporarily.

Throughout the tour, Mark Smyk, regional resident geologist for MNDM, and Gerry White had been asking questions, making observations, and recording meticulous notes.

Earlier in the season, a Sage crew had stripped the showing for 12 metres and extracted samples from 12 channels, spaced a metre apart. Bill Love explained that the exposed vein at that time averaged 19.5 g/t, with up to 9 g/t found in the wall or host rock. Ted Cox stated that another 34 samples have been sent for assay, and that they were preparing more to go.

Last winter Sage Gold had run geophysics on the Jacobus East Area, and when the snow had gone, Nolan Cox had been stripping a "line feature" farther west without getting any values. Then in July a prospecting team found a patch of bedrock, about 2 by 3 metres in size, which is now called the Golden Extension Showing.

One of the investors, J.C. St-Amour of Frazer Mackenzie, told the writer that he had been impressed by the earlier presentation. That the Sage find was on line of strike with Kodiak's Golden Mile looked very promising. As for the Golden Extension, he said, "This is still early [in the operations stage], and it will be good to see some drill results."

Another interesting feature of the showing was that someone had used a spray can to print on top of the dome in the trench the following words: GOLDEN SPIDER. Nolan Cox displayed a photo of a yellow spider whose home they had disturbed. Until the showing received its official name, the crew had apparently been using the term as a workplace title.

Back at Pasha Lake Cabins, the party relaxed with beverages in hand as they hovered over display boxes of split core and a table loaded with interesting samples from other Sage properties in the Jellicoe area. The day ended with a lavish meal catered by Karen Ostler of Geraldton.

The following day, Tuesday, the tour would move to Sage's Onaman Property.

Part II – The Onaman Property

Very few people have ever visited the oldest gold mine in the Beardmore-Geraldton mining camp. In 1928, Tashota Gold Mines Ltd. began sinking a shaft 16 miles south of the northern line of the CNR, in the Onaman Lake area.

On May 3, 2006, Sage Gold Inc. optioned the Onaman Property from two unrelated individuals, and now has 100% ownership of claims with significant values for base and precious metals. The old mine is known colloquially as the Tashota-Nipigon,

On Tuesday, September 23, 2008, the writer accompanied a field party touring the Onaman Property. The day started cool and overcast. The four 4-WD pickup trucks turned east around the Kilometre 54 marker of the Kinghorn Road, north of Jellicoe, and traveled a good single-lane road for 3.6 kilometres (km). The party disembarked to examine a huge overburden trench, also described as an open pit. The pit measured perhaps 90 metres long, and 30 metres (m) at its widest point.

On September 23, a field party on the Onaman Property clamber about the open pit of the Lynx North Zone.

The party - comprised of Sage officials, investors from Toronto, and two government geologists - examined the rusty bedrock, and listened to a presentation by Bill Love, Vice President/Business Development, and Nigel Lees, President. The "rust" or gossan, according to Ron Therriault, project geologist, came from the weathering of sulphide minerals in chalcopyrite, pyrite, and pyrrhotite. Other important minerals were ankerite, argentite, and copper bloom (malachite, a bright green colour).

The night before, back at Pasha Lake Cabins on the 801 Road, the party had lingered over displays of drill core and grab samples from Sage's properties. One spectacular split core, from Hole S08-52, gleaming like a brass pipe, had returned assays of 8.07% copper, 6.08 grams per tonne (g/t) gold, and 236.0 g/t silver, over a length of 3.72 m.

Recently Sage has determined that three polymetallic zones, formerly dubbed Lynx 1 (now Lynx South), Lynx 2 (North), and Lynx 3 (Centre), are really one system. Other showings occur over a stretch of about 10 km from the old Tashota-Nipigon mine in the north to Km 50 in the south.

Sage was especially interested in two other zones to the west of the Lynx system. The Headway Main Zone is zinc-rich, and the Big M has copper, gold, and silver values. So far the company has conducted minimal exploration on these zones, but, according to Love (VP), Sage will bring in experts next summer to determine if these two zones and the Lynx system are interrelated.

"The best developed [explored] of all the copper zones is the Lynx system on the Onaman Property," said Love. Sage has data from 106 drill holes over a strike length of 1 km, and to a depth of up to 200 m, for a total drilled length of 19,941 m. The system is "open" (that is, has mineral potential) at both ends and below 200 m. Nineteen of the holes are historic (i.e., predating Sage's involvement). Sage has directed a "QP" (Qualified Person) to provide numbers compliant with the government-mandated regulation called NI 43-101. An independent examiner will then calculate the size of the mineral resource by the end of this year.

Commenting on the geology of the trench (in Lynx North), Lees (President) said, "It's probably a good place to run courses on geology." He told the writer that the Lynx zone system may be compatible with an open pit mining operation.

A prominent landmark of the old Tashota-Nipigon mine is the waste rock dump.

The field party then proceeded along the narrow road for another 5 km, crossing a plank bridge over a stream. The trucks stopped with a tailings area (what local residents call "slimes") on the left and the concrete foundations of an ancient mill on the right. Walking back along the road a few metres, the party took an overgrown tote road running east, and in about 90 m, they encountered a clearing of sorts. They had arrived at the old Tashota-Nipigon.

Gerry White, MNDM, records the exact location of the old shaft with a GPS unit. Lyle Holt looks on.

A big mound of waste rock dominated the clearing. There were evidences of wooden buildings here and there, and the remnants of concrete structures. Myron Nelson, prospector, immediately began tapping rocks with his hammer. His colleague, Lyle Holt, directed the group to the mine shaft, capped with concrete and measuring about 4 by 7 feet. No one, wisely, accepted a dare to stand on the crumbling concrete. Until a few years ago, it was not unusual to find shafts open to the sky in this mining camp, and some old shafts have yet to be rediscovered.

In the 1930's, miners had accessed this site from the railway or by airplane. There was a winter road from Tashota, a whistle-stop on the CNR, and in summer, an 18-mile canoe route. There was a 6-mile wagon road from Onaman Lake.

The tour of the Onaman Property ended, the party headed out. The sun peeped through the clouds, lifting everyone's spirit just a tad. Next stop was a gold property back on the 801 Road, Sage's Paint Lake Property.

Later, one of the investors remarked to the writer that the old adage holds true: "The best place to find a mine is in the shadow of a headframe." Whether that headframe still stands or not. Donna McPherson is a stock broker associated with Octagon Capital Corporation.

Readers interested in values from channel or drill core samples from any of Sage's properties, check out news releases at www.sagegoldinc.com.

Part III – PAINT LAKE & MISSING LINK

Sage Gold Inc. has acquired a property next door to the largest published gold deposit in the Beardmore-Geraldton mining camp. The Paint Lake Property is along strike with the Brookbank Gold Deposit of Ontex Resources Ltd.

On Tuesday, September 23, Sage's investors field tour arrived at the property just before 1:00 p.m. The field party turned west on a logging road off the 801 Road (named Hwy. 801 at this point), 8.4 kilometers (km) north of Hwy. 11. The road soon deteriorated into a trail recently pushed through the bush. In just over a kilometre, the four vehicles could proceed no further.

Sage officials broke out the picnic lunch. They passed around water, sandwiches, and cookies. Individuals found seats on a small bedrock exposure, or walked about examining the curious structure. Ron Therriault, project geologist, described the rock as mafic volcanic pillows and tuffs. For the technical enthusiasts, he described the host rock as crosscut by quartz-ankerite-pyrite veins, bordered by chlorite-sericite-silica-pyrite altered and stockworked selvedges.

Unlike the Golden Extension Showing at Jacobus East, no single vein dominates the structure. The host rock is chockablock with quartz-ankerite-pyrite veinlets. Munching their sandwiches or cookies, the party ambled a few metres further west to the main trenching operation. Sage has identified 300 metres (m) of mineralization to date, with widths up to 3 m on individual vein-selvedge packages. Nobody was working the property that day. From the south came the machine noises of a forest harvesting operation.

Bill Love, Vice President/Business Development, stated that Sage had leased the 6 claims from Nolan Cox, who had also done the stripping. To the west about 2.5 km away, beyond a barrier of bush, lay the Brookbank deposit, compliant with National Instrument 43-101, one of the key regulations governing the mining industry.

The Ontex Resources property spans a 35-km strike length of favourable geology and structures with several important gold zones including the Brookbank. The Brookbank already has a proven and estimated resource of 694,700 ounces of gold in ore grading 2.0 grams per tonne (g/t) and better. In that respect, Ontex is more advanced than the mining camp's other leading player, Kodiak Exploration Ltd.

Ontex has an ongoing drilling program on or near the Brookbank Gold Deposit. Exploration is proceeding in or near the Paint Lake Deformation Zone (aka the Shear Zone or the Fault), which trends generally east-west. To the east of Brookbank, the Paint Lake Deformation Zone remains a prominent feature of Sage's Paint Lake Property.

On the Paint Lake Property, Bill Love (2nd from L) leans down to examine a structure. David Lees, cameraman, records the moment for a corporate video. Nigel Lees (2nd from R) converses with an investor.

On September 23, the visitors scattered over the rusty bedrock. The prospectors and geologists got down to business, tapping rocks, and initiating intense discussions with their colleagues. Other visitors drifted in and out of the discussions.

Asked what he thought of the two-day tour, Craig Stanley replied that he looked forward to getting more information. Stanley, mining analyst for Blackmont Capital, based in Toronto, said he would really like to see drilling results (on properties such as Paint Lake and Golden

Extension) and/or a proven ore body. However, he said, "The fact that you can drive to these places is amazing."

Mark Smyk, MNDM, sweeps dirt off the conglomerate outcrop as Sage officials look on.

All four investors were impressed by the infrastructure in place or at hand: the road system, the energy sources (hydro and natural gas), and established communities for services, accommodations, and work force.

On August 25, Sage had published channel sample results. Of the 14 best results, the lowest value was 0.39 g/t over 0.7 m; the highest, 23.09 g/t over 0.4 m. Eight samples graded higher than 2.47 g/t.

Returning along the trail, the party stopped to examine some outcrops of conglomerate. The MNDM geologists, Mark Smyk and Gerry White, kneeled down and eyeballed the rock with obvious interest. Apparently the Brookbank deposit is also bordered on the north by the same formation.

The field party returned to Pasha Lake Cabins for an hour of relaxation and conversation. The chartered bus having arrived, several visitors left around 3:30 p.m. to catch a plane to Toronto from Thunder Bay Airport.

In the interval, the writer interviewed Love (VP) and Nigel Lees, President and CEO. Lees stated that Sage Gold believed in a proactive approach in its relationships with First Nations (FN). He pointed to the Memorandum of Understanding which Sage had signed with a FN near Timmins with regard to Sage's Kerrs Property. Recently he had given a presentation to AZA (Animbiigoo Zaagi'igan Anishinaabek), based in Beardmore.

In response to questions, Lees stressed that junior mining companies such as Sage Gold do not generate revenues. They must raise money to finance their operations. Sage can and does advise FN of job opportunities, and promotes training programs, he said. Most jobs, of course, lie with the contractors that Sage hires (staking, prospecting, line-cutting, geophysical work, etc.).

Lees was very pleased about the infrastructure and the mining potential of the area. "We want to put our time and effort into developing our Beardmore properties," he said. He noted the "cheap" exploration costs associated with a work force that goes home after its shifts.

A few days later, the writer accompanied Ron Therriault to Sage's Missing Link Property. On September 30, the party left their vehicles 7.5 km up the Kinghorn Road, and walked east on a logging road for over 300 m. Then they struck out south along a footpath for about 45 m to a clearing that ran generally east and west.

On the Missing Link Property, Ron Therriault, project geologist, places his foot on an historic channel cut. In front of him the bedrock has been pitted to some extent by oldtime prospectors.

Years ago, Nolan Cox had stripped the shallow overburden on this extension of the Paint Lake Deformation Zone. Grass and moss and scattered saplings were attempting to reclaim the area for nature. "We are looking," said Therriault, "at strongly deformed and sheared mafic volcanics cut by quartz ankerite arsenopyrite veins."

As in the Paint Lake Property, prominent veins are few, and pinch and swell along strike; others are very small, even microscopic. "Many of the veins tend to be discordant to the main fabric [foliation] of the shear zone," said Therriault. One can also observe historic channel sample cuts and some shallow pits.

As for the Missing Link Showing, "Right now it's not a top priority [for Sage]," said Therriault. "Sage is more interested in Solomon's Pillars, given the success of Kodiak's drilling at [Sage's] boundary, and [the values in] historic samples."

The Solomon's Pillars Property is accessed from the 801 Road, and to date, as for the Missing Link, only grab samples have been sent away for assay.

In concluding this series, let us hark back to a remark by Nigel Lees in his initial address to the field party: "We think [the Beardmore-Geraldton area] is underexplored, has great potential, and we've got great hopes here."

- Times Star

SAGE FOCUSES ON BEARDMORE-GERALDTON GOLD CAMP

"We really want to concentrate on the Beardmore-Geraldton area. We think it is underexplored, has great potential, and we've got great hopes here."

Nigel Lees, President and CEO of Sage Gold Inc., was speaking to the writer, a group of investors, and two geologists from the Ministry of Northern Development and Mines (MNDM).

Sage has several gold properties - the Kerrs Property near Timmins, and others in Nevada and Arizona. When Sage began exploring in the Beardmore-Geraldton mining camp in May, 2006, it targeted two properties with historical resources. It acquired the Jacobus property, up the 801 Road, west of Jellicoe, and the Onaman Property, up the Kinghorn Road, east of Jellicoe. Historically, they had produced good values in copper, nickel, and zinc, along with some gold and silver.

Nigel Lees, President & CEO, perches on a pile of blasted rock as he points to a map of Sage's Lynx Zone system. To his left is a sampling channel which was cut into this open pit area.

Tuesday, September 23, found Lees standing in an elongated exploration pit west of Onaman Lake on the best copper-rich showing of the Lynx Zone system. Back at Sage's field base the night before, the party had examined one spectacular core from Hole S08-52. Over 3.72 metres, the core returned values of 8.07% copper, 6.08 grams per tonne (g/t) gold, and 236.0 g/t silver. Other results can be obtained from news releases beginning on May 28, 2008 on Sage's website: www.sagegoldinc.com.

The party touring Sage's properties was standing on very rusty bedrock that had been cut for channel samples, blasted for bulk samples, and drilled for core samples. A convoy of four 4-WD pickup trucks had traveled 3.6 kilometres along a well-maintained single-lane bush road from the Kinghorn Road, starting near the Kilometre 54 marker.

Lees explained that the three Lynx Zones (1, 2, and 3) have proven to be one system, stretching along a north-south strike length of one kilometre. In the vicinity of this massive to semi-massive sulphide zone are two other sulphide showings. The Headway Main Area is zinc-rich and the Big M has "spectacular alteration" with copper, gold, and silver values. There are several other showings that follow a geophysical anomaly stretching from the old Tashota-Nipigon mine in the north to the Km 50 showing on the Kinghorn Road in the south, a distance of about 10 kilometers.

Bill Love, Vice President/Business Development, stated that Sage has data from 106 drill holes, - some of them historic holes - for the Lynx system, for a drilled length of 19,941 metres. Mineralization is open to the north and south, and open below a depth of 200 metres. The most recent drill holes yielded values from 0.75% to 11.36% for copper, 0.10 to 2.64 g/t for gold, 18 to 264 g/t for silver.

The numbers are now under review by Peter Hubacheck, consulting geologist, to achieve compliance with NI 43-101, a requirement under Ontario's mining regulations. Very shortly the data will be turned over to an external, independent examiner in order to calculate the size of the mineral resource on Lynx (i.e., the ore body, if such exists at this point).

Love also stated that Sage suspects that Headway Main Zone and Big M, on both of which Sage has done minimal exploration, will prove to have a stratigraphic connection to the Lynx system. To that end, Sage plans to bring in several experts next summer to explore that possibility. "How the three components interrelate is the big question," he said.

The Jacobus deposit up the 801 Road, now known as the Jacobus South Zone, is copper-and-nickel-rich. The day before, the party had visited a new discovery in the Jacobus East Area, called the Golden Extension. Sage has made what it believes will become a major gold find.

Back at the Lynx pit, Lees told the writer that Lynx would lend itself to an open pit mining operation. The Onaman Property deposits are "polymetallic". The high price of precious metals in the today's market, said Lees, is persuading the company to concentrate more on its many gold properties in the Jellicoe area.

Ron Therriault, project geologist, explained that the "rust" or gossan comes from sulphides (chalcopyrite and pyrite). Other main minerals are pyrrhotite, ankerite, argentite, and malachite.

Before visiting Sage's Golden Extension Showing the day before, the field party learned that geologists from both Sage and Kodiak Exploration Ltd. have a strong belief that the showing is an extension of the huge gold finds on Kodiak's Hercules Property. The party had first visited Kodiak's Golden Mile Zone, where two top Kodiak executives outlined what work they had done and what they proposed to do.

On the Golden Extension Showing, Ted Cox, field geotechnician, shows investors a gold tail obtained by panning soil near the vein.

Gerry White, district geologist for MNDM, advised the writer that if there were a confirmation of an extended strike length of the gold-being vein system, there were opportunities for additional significant gold discoveries by both companies.

Sage's prospectors found the Golden Extension Showing only this past spring, in an area contiguous to Kodiak's Hercules Property. Hampered previously by a lack of road access, Sage had been machine-stripping the broad quartz vein for only a week, confirming 200 metres of strike length in the same northwest trend of the Hercules veins, and in the same host rock.

On September 23, heading back to the field office, the party stopped at Sage's Paint Lake Property just off the 801 Road. Located in the Paint Lake Shear Zone fault, a very important structure, the property abuts the Brookbank gold property of Ontex Resources Ltd. to the west.

Work on the Paint Lake Property has barely begun. Therriault said that so far Sage has grab and channel samples. Of 14 channels for which there are published results, 7 have values ranging from 4.04 to 6.44 g/t gold, and one is 23.09 g/t over a width of 0.4 metres.

The Brookbank Gold Deposit has an estimated resource of 629,900 ounces, and drilling continues on that property.

During the two-day tour, two government officials – Mark Smyk, regional resident geologist, and Gerry White – recorded their observations.

Three geologists on the Paint Lake Property hold a technical discussion. Left to right, Mark Smyk of MNDM; Bill Love, Sage VP; and Ron Therriault, project geologist.

Other properties that Sage Gold is actively exploring for gold are Missing Link and Solomon's Pillars, with some other prospects still to be explored.

A year ago, Sage stock reached a one-day high of 80 cents a share; on September 23, it closed at 12 cents on the Toronto Ventures Exchange.

It appears that Sage Gold Inc., as far as its Jellicoe properties are concerned, is now determined to live up to the metal in its name.

- The Chronicle-Journal, under the heading "Getting a Prospector's Perspective"

BEYOND THE GOLDEN MILE

Bill Love's six-foot-four frame straddles a quartz vein as wide as his hips. It is a sunny autumn day in the middle of the northern bush east of Lake Nipigon.

For the past three years, Sage Gold Inc. (SGX:TSX-V) has been exploring its base metal properties in the Beardmore-Geraldton mining camp. Now that gold is around USD $900 an ounce, Sage is growing more excited about the gold values in its assays.

Bill Love, Vice President/Business Development, is standing on Sage's newly discovered gold property in the Jacobus East Area. Last fall and winter, they were focusing on the Jacobus South Zone and the Onaman Property, pursuing values in copper, nickel, and zinc, mingled with interesting gold and silver numbers. This spring they sent out prospectors to scour their property adjoining the huge gold find at the Hercules Property of Kodiak Exploration Limited.

Their prospectors struck gold in an outcrop about 5 kilometres northwest of Kodiak's Golden Mile Zone. Located in a clearcut in the Lake Nipigon Forest, the new showing had a bush road to it, but the road was impassable because the Ministry of Natural Resources had pulled a bridge over Fairview Creek.

That sunny day, Monday, September 22, Nigel Lees had led the writer, a group of investors, and a number of geologists to the find they call the Golden Extension Showing. Lees, who is Sage's President and CEO, led the party from the field office, located between the small communities of Beardmore and Jellicoe, which are accessed by Highway 11.

The Kinghorn Road is a primary bush road striking north of the highway. Just past Kilometre 43 on the Kinghorn, a secondary road heads west for almost 11 kilometres, over the newly installed bridge, to the Golden Extension.

A backhoe operator is engaged in stripping the overburden, and a group of men are washing down the vein and taking channel samples, which they extract with a special rock-cutting saw. The machine has been operating only a week, and already 200 metres of strike length have been confirmed although not completely trenched.

Up to a week ago, Sage's crew had been stripping the outcrop by hand, and had exposed 12 metres. Initial channel samples had been positive. Seven of the 12 channels yielded values ranging from 2.08 to

81.1 grams per tonne (g/t). An eighth sample (CH-05C-10) assayed at 136.5 g/t over a width of 0.50 metres.

Bill Love, VP/Business Development, examines the visible gold in a channel sample on the Golden Extension. He then encourages John Walter, an investor relations consultant, to take the rock home. That may explain Walter's smile.

The broad quartz vein runs through stock known as the Elmhirst Lake Intrusion, a granodioritic rock. Ted Cox, Sage's field geotechnician, pans a handful of topsoil for the visitors, and points out the gleaming gold tail. He has also found several instances of visible gold (VG) in the vein and gold values in the host rock. The vein at its broadest point, he says, is 1.3 metres. He recently sent out 34 channel samples for assay, and is preparing another batch.

Sage has provided several channel samples for the visitors to cart away. It takes an expert eye to spot the VG. Cox describes one sample: quartz, 20% pyrite "married" to VG, 2% chalcopyrite, and black streaks which are magnetite.

Earlier in the trip, the Sage party had stopped at Kodiak's Golden Mile Zone, smack in the middle of its "jewelry box" where the quartz

A geotech hoses down the broad quartz vein to bring out the mineral colours, including microscopic specks of free gold.

vein is 3 metres wide. There, two of Kodiak's top executives had laid out their progress to date: 3 kilometres of strike length, most of it stripped and all of it drilled, trending to the northwest.

Aerial view looking generally north over the Golden Extension trench on September 22. In the upper left, out of view, the trenching continues. MNDM Photo.

Kodiak has found some spectacular gold values not only in the veins but in the host rock itself, the Elmhirst Lake Intrusion. Their drill hole intersections dip down to 550 and 570 metres already. They have stripped and drilled two parallel zones. They send about 1,000 samples a day to the assay lab.

Standing in the "jewerly box" of the Golden Mile Zone, a mixed group, which includes investors, receives an update from Brian Maher, Kodiak's VP/Exploration. Nigel Lees, Sage's President, looks on (upper left).

One of Sage's visitors comments: "I'm speechless. It's got tremendous strike and breadth and multiple vein structures with the opportunity [for all of them] to meet down dip." The result could very well be an "economic resource", i.e., an ore body to justify a mine. John Walter, the speaker, runs an investor relations firm on Bay Street.

This news is good for Sage Gold Inc. The writer and two others climb into a Bell Long Ranger helicopter and fly northwest along the Golden Mile line of strike. In a few minutes, the party disembarks at Sage's Golden Extension.

The zone is aptly named. None of the experts present doubts that the Golden Extension is indeed an extension of a vein system on Kodiak's Hercules Project.

Since that day in September, prospectors have identified a new vein near the boundary with Kodiak, along strike with Sage's Golden Extension and Kodiak's Golden Mile. Additional veins have been stripped in the vicinity of the original showing (Vein #1), for an aggregate strike length of 290 m in early November.

According to Nigel Lees, President, the company is currently devoting significant resources to its known gold holdings around Beardmore/Jellicoe, which include the properties of Paint Lake, Missing Link, and Solomon's Pillars. Other prospects have yet to be explored.

The company has gold properties in Kerrs Township, near Timmins, and in Nevada and Arizona. On September 22, Sage Gold shares closed at 12 cents, although its stock reached a one-day high of 80 cents last October.

- *Northern Ontario Business*

SAGE GOLD INC.

Precious and Base Metals in the Americas

TSX V: SGX

SAGE GOLD INC.
365 Bay Street, Suite 500
Toronto, Ontario
M5H 2V1
Tel: (416) 204-3170
Fax: (416) 367-2243
info@sagegoldinc.com
www.sagegoldinc.com

FOCUS ON BEARDMORE-GERALDTON GOLD CAMP

2008 DISCOVERY OF THE "GOLDEN EXTENSION" ON JACOBUS EAST PROPERTY

- Golden Extension believed to be the northwest extension of Kodiak Exploration's Golden Mile
- Discovered twelve new veins with an aggregate trenched strike length of over 1 kilometre
- Best Channel sample to date of 136.5 g Au/T (3.98 oz/T)
- New channel results up to 80.4 g Au/T over a channel width of 0.5 metres.
- Trenching and Geophysical Exploration ongoing.
- Veins open along strike
- Drilling planned for spring/summer of 2009 on Golden Extension
- NI43-101 Resource Estimate pending on the Lynx Zone at Onaman property

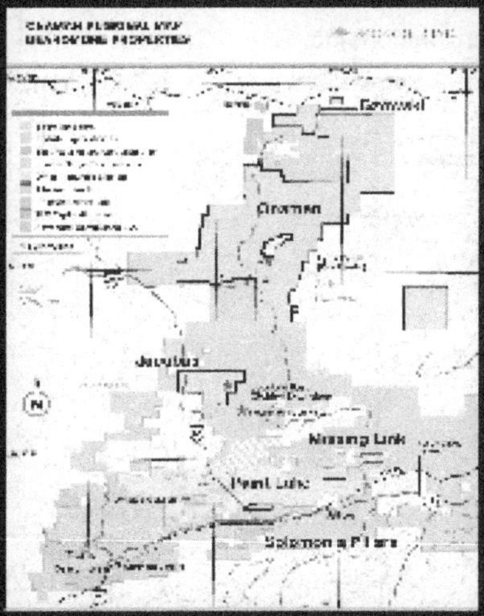

Golden Extension Vein 1-5

Visible Gold Sample for Vein 4 of Golden Extension

SAGE GOLD INC.

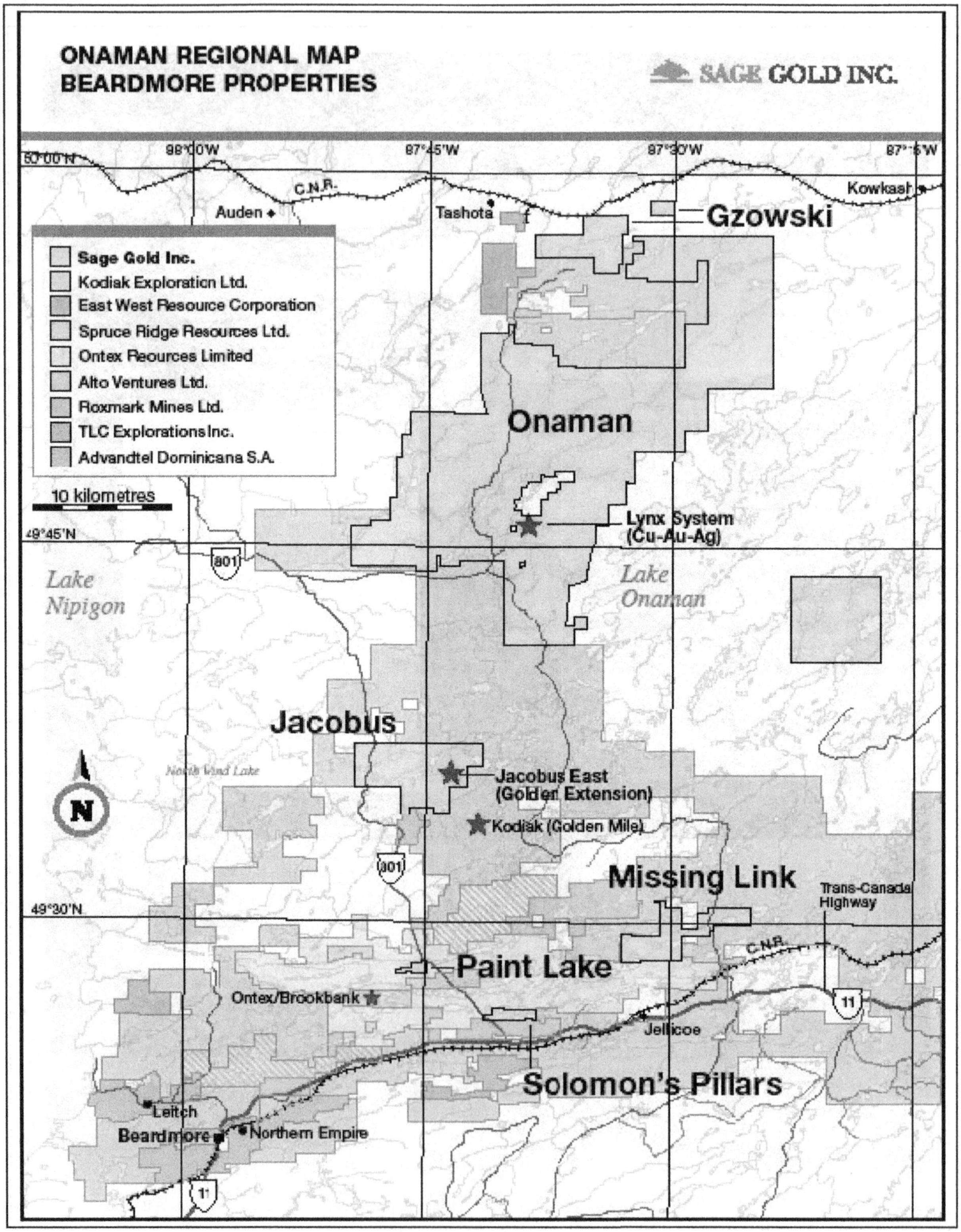
ONAMAN REGIONAL MAP
BEARDMORE PROPERTIES
SAGE GOLD INC.
50°00'N
88°00'W
87°45'W
87°30'W
C.N.R.
Auden
Tashota
Kowkash
Gzowski
Sage Gold Inc.
Kodiak Exploration Ltd.
East West Resource Corporation
Spruce Ridge Resources Ltd.
Ontex Reources Limited
Alto Ventures Ltd.
Roxmark Mines Ltd.
TLC Explorations Inc.
Advandtel Dominicana S.A.
Onaman
10 kilometres
Lynx System
(Cu-Au-Ag)
49°45'N
801
Lake
Nipigon
Lake
Onaman
Jacobus
Jacobus East
(Golden Extension)
Kodiak (Golden Mile)
N
801
Missing Link
Trans-Canada
Highway
49°30'N
C.N.R.
Paint Lake
Ontex/Brookbank
11
Jellicoe
Solomon's Pillars
Leitch
Beardmore
Northern Empire
11

Scenes from Sage's
Golden Extension Showing

Chapter 4 :
Year 2008

Activities of Premier Gold Mines Limited

Muskeg Tours 2009

PREMIER
Gold Mines Limited

One challenging site for the drillers, May 27th

PREMIER GOLD REPORTS

The Executive Vice President & Chief Operating Officer for Premier Gold Mines Limited presented an overview of his company's activity. Stephen McGibbon, P.Geo., concluded with a look at its Geraldton Project.

McGibbon was addressing the delegates to the Northwestern Ontario Associated Chambers of Commerce. After the presentation, he confirmed details with the Times Star.

Stephen McGibbon of Premier Gold visits one of the two drilling rigs operating on the old Little Long Lac mine property.

Premier Gold has had a long association with the Red Lake area. Other current projects include one near Pickle Lake and one in Mexico. McGibbon explained that his company believes in exploring multiple properties to reduce risk to investors.

It is the policy of Premier Gold, he said, to target areas that have a past record of successful mining and exploration. "Typically," he said, "these districts are high-grade districts." The company believes that the Geraldton area has been underexplored, and that there is a high potential for more major discoveries.

He pointed out that other gold camps in Northern Ontario have explored to great depths. Kirkland Lake mines dipped to 8,100 feet below surface; Red Lake mines to 6,700 feet. "In Geraldton, past production and exploration was generally limited to 2,500 feet below the surface."

Recent drilling near the old Little Long Lac mine has been quite shallow, within a couple of hundred metres of surface. Premier, he said, is looking at the obvious, high potential targets now, and if the values are good, the company will explore at greater depth next year.

Premier has run geophysical surveys of the old Magnet mine, the TAZ area, and the old Little Long Lac mine, and is targeting certain rock units. "What is a very common model for gold exploration in areas such as Geraldton," he said, "is looking for rock units that have been folded and faulted and altered and disrupted, because ultimately it is those cracks in the rocks that fluids containing gold percolate within and deposit the veins."

Two drills have been operating almost continuously since last October. McGibbon expected that a third drill would begin work in June. Since October, the company had spent between 1.5 and $2 million, and in this calendar year, 2008, Premier expects to crank it up. "We're going to be on track for spending between 4 and $6 million in 2008." Each metre of drill core costs Premier about $125.

Earlier in April, the company released the results of the first two holes in the newly discovered Kailey Zone at Little Long Lac. PLL08-009 returned an intersection of 51 metres grading 1.33 grams per tonne (g/t) gold, and PLL08-012 returned 26 metres grading 1.34 g/t.

The holes were shallow, and the grades were low, but combining these results with other information, Premier announced that there was potential for an open-pit style mine. McGibbon expressed the hope that sometime in 2009 the company could disclose a 43-101-compliant mineral resource.

At some point Premier would be targeting a zone of sedimentary rock under Barton Bay, east of the causeway, just off the north shore. "[This] arkose unit," he said, "was the host for mineralization for Little Long Lac mine in the past." Like many other targets in the area, it was never explored. A strong fault offset this unit from the structure that lies under the old mine.

Asked if Premier Gold would have a phase of stripping and trenching, McGibbon replied, "The nuts and bolts of our exploration is going to continue to be diamond drilling." He said there were very few outcrops in the Geraldton area which would indicate "thin" overburden. The company believes that, although it is more expensive, drilling is the quickest way to delineate a deposit. However, McGibbon did not rule out surface exploration if drill results are good and soil is shallow.

How long would Premier be exploring in the Geraldton area? McGibbon said that President Ewan Downie had researched the area for a long time before Premier optioned property from Roxmark Mines Limited.

He emphasized how serious Premier is: "Our commitment to Geraldton is a long-term commitment."

- *Times Star*

THE MINING CAMP AWAKENS

"This is a sleeper camp. I can't believe exploration died out here in the '50's."

Tim Twomey, exploration manager for Premier Gold Mines Ltd., was addressing GEDC's Mining Procurement Workshop, Part II, on Wednesday, October 22.

Earlier in the day, Premier had announced an agreement to purchase mining claims held by Lac Properties, Inc., a subsidiary of Barrick Gold Corp. The claims, known by Barrick as the Geraldton, Ozone Creek, and Eva Summers properties, encompass the former producing mines of McLeod-Cockshutt, Hardrock, and Mosher.

Premier believes the G-L Property, as it calls the claim group*, still has undiscovered reserves of gold in the old mines which produced 2.146 million ounces between 1934 and 1968.

When the last mine - the McLeod-Cockshutt/Mosher - closed, miners had been developing the F Zone at the 2,621-foot level. Twomey pointed out that the average depth of the old mines here "is pretty shallow for greenstone archean mines in Canada." One slide in his presentation referred to the maximum depths on mines in Northern Ontario gold camps; e.g., Kirkland Lake, 8,075 ft.; Timmins, 7,200; Red Lake, 6,750; Hemlo, 4,000; and Geraldton, around 2,400.

Two mined zones, the F Zone and the North Zone, remain open at depth, meaning there is a potential gold resource at greater depth. These zones are aligned east-west, to the north of Hwy. 11. One drill intersection in the F Zone (in the west end) returned values of 13.75 grams per tonne (g/t) over 6.4 metres (m) down-dip (beneath the mine workings). "They developed all this ore," he said, pointing to a slide, "and shut it down." In the North Zone (further east, nearer the headframe), one hole assayed at 18.41 g/t over 5.1 m down-dip.

The historical resource of the property is 3,000,000 short tons valued at 0.08 g/t. By law, of course, this resource must now be verified using new yardsticks – or metresticks, as the case may be.

Dina Quenneville, executive director of GEDC, is captured conversing with Tim Twomey of Premier Gold at the Prospects North mining conference in Thunder Bay on October 24.

The G-L Property has other potential: (a) There are significant resources reported to remain in the old mine workings. (b) Several newer high-grade discoveries which had been made in other areas have never been followed up. (c) In the area of the glory hole and Porphyry Hill (east of the headframe), there is a reported resource of several hundred thousand ounces that would be compatible with an open pit operation.

Premier expects to begin a diamond drilling program in the new year. In the meantime, it will carry on with exploration of its Geraldton Property, in joint venture with Roxmark Mines Ltd. The Geraldton Property includes the old mines of Little Long Lac, Magnet, and Bankfield.

Twomey also pointed out the advantages enjoyed by both major properties in this exploration phase, such as an available work force, a fully serviced community, and a major highway connection. "I'm really impressed with the welcome I get here in Geraldton," he said. "I've never seen infrastructure like this," he added.

In a more prosperous past, the McLeod-Cockshutt mine dominates the landscape. The original headframe (centre, near the junction) still stands today. In the upper left is the Mosher mine's headframe and water tower.

Twomey, who has 25 years of experience in the mining sector, joined Premier recently after leaving Goldcorp Inc., Red Lake's major employer. He had played a key role in the discovery, development, and production phases of the High Grade Zone of the Red Lake Mine, Canada's largest gold mine and one of the world's richest.

The deal with Barrick Gold means Premier will acquire a 100% interest in the G-L Property, making a $1 million cash payment and turning over 500,000 shares. Premier must also deposit $1 million in a reclamation trust fund, and grant a 3% net smelter return royalty (in the event a producing mine develops).

Premier must also transfer an interest in the property to its joint venture partner, Roxmark.

Premier Gold Mines Ltd. also operates the Red Lake Project (in joint venture with Goldcorp), the PQ North Project (near Goldcorp's Musselwhite mine, which has historically produced over 2,000,000 ounces of gold), and the Santa Teresa Project in Mexico.

Twomey said that Premier wants to recover old mining records from the Geraldton area, specifically digitized data, which he believes might be stored in someone's basement and forgotten. Anyone with information should contact Premier or the Times Star.

Twomey concluded the presentation with this remark: "We really have some [terrific] real estate here for exploration."

On Friday, October 24, Premier's shares (PG) closed on the TSX at $1.55.

- Times Star

*The G-L Property is now referred to as the Hardrock Project.

Refer to page 178 for the Chibougamau Drilling Ltd. ad.

PREMIER
Gold Mines Limited

PREMIER GOLD MINES LIMITED – Hardrock Project, Ontario

Ewan Downie President and CEO proudly stands in front of the Macleod-Cockshutt headframe located on the Hardrock property next to the trans-Canada Highway.

Premier Gold Mines Limited ("Premier") is a fully financed company, focused on the development of several highly prospective gold projects in Canada. In addition to the advanced-stage Red Lake (Goldcorp Joint Venture) Project, Premier has recently announced the acquisition of Barrick Gold's properties in the prolific Beardmore-Geraldton gold mining district. The "Hardrock Project," is located in the Geraldton region of northwestern Ontario, Canada. When combined with the existing Geraldton Project, which Premier is currently earning a 70% interest from Roxmark Mines Limited, Premier now controls more than a 15 km strike length of the most prospective, and under-explored, geology in the district. This Property has had past production of nearly 3 million ounces of gold and is wide open for exploration.

The Hardrock Project represents one of Canada's premier near-term development projects with multi-million ounce gold potential from both open pit and underground zones. Until the mid-1970's, this highly prospective property has seen nearly 3 Million ounces of gold production from operations limited to depths of only 600 meters. Since this time, essentially no deep exploration has been performed over the core area for the past several decades, making Geraldton one of Ontario's most under-explored major gold belts. The property can be staged, with immediate resource definition and expansion drilling, then potential near-term mine development from several known deposits on the Project.

The drills started turning mid January. In the distance, the Macleod- Cockshutt headframe, the Geraldton Interpretive Centre and the Trans Canada Highway.

Known (historical) resources exist within open pit and underground zones located proximal to and within the historic mine workings. While all production on the project area came from underground operations, more recent work has focused on delineating open pit resources including at the recently discovered "Kailey Zone". There are numerous high grade vein prospects exist throughout the project area in addition to several untested targets located proximal to the historic operations.

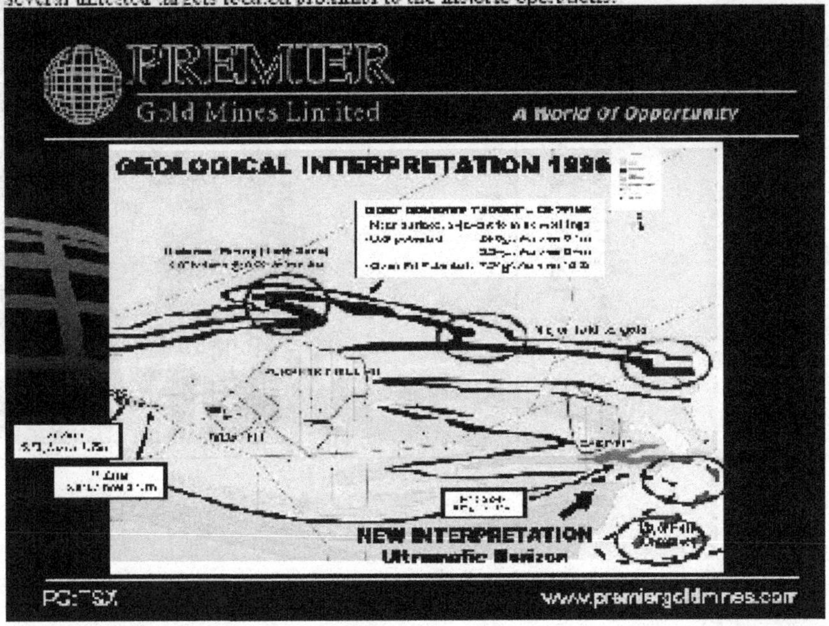

Above is a geological interpretation of our Hardrock Project. Notice the numerous untested areas highly prospective areas.

Premier plans to spend an estimated $10 million dollars and drill up to 65,000 meters on this project alone in 2009. The focus of this project will be to gain sufficient information to release a NI 43-101 compliant resource of the historic workings as well as other known deposits by year end. Premier will also be testing numerous highly prospective and unexplored targets on this ground. For further information about the Company, please feel free to contact us or visit our website.

For further information, please contact:
Ewan Downie, President & CEO   Phone: 807-346-1390   Fax: 807-346-0100
E-mail: info@premiergoldmines.com   Web Site: www.premiergoldmines.com

The statements made in this document may contain forward-looking statements that may involve a number of risks and uncertainties. Actual events or results could differ materially from the Company's expectations and projections.

Chapter 5 : Year 2008

Activities of Noront Resources Limited & Spider Resources Inc.

Muskeg Tours 2009

The Attawapiskat River winds through the James Bay Lowlands on March 26th

RING OF FIRE IN JAMES BAY LOWLANDS

The man with a dream sits on a wooden bench 190 kilometers north of the Albany River and 240 kilometers west of James Bay.

Neil Novak, P.Geo., is wearing two hats today, the real one of which reads Spider Resources Inc., and the other is the shirt he is wearing, which spells out Noront Resources Ltd. Novak is President and C.E.O. of Spider (SPQ:TSX-V) as well as Vice President/Exploration of Noront (NOT:TSX-V).

The kitchen tent is not a quiet place this morning. The two cooks clatter about. Men bundled up against the minus 17 degrees C. stomp in and out. They relish the respite from a stiff wind out of the northwest. They stomp around on the wooden floor in insulated rubber boots as they grab a coffee or sit down for a sandwich.

It is Wednesday, March 26, 2008, and a Cessna Grand Caravan, courtesy of Billiken Management Services Inc., has just shuttled the writer and the photographer Jim Guillemette from Nakina airport – the end of the paved road – to an ice strip on McFauld's Lake in the subarctic.

About a dozen and a half tents squatting on the frozen muskeg constitute the Billiken Base Camp. A few times a day a Caravan and occasionally a Hawker Siddeley 748 drops off another payload of supplies and equipment to run the camp and to sustain crews farther afield.

Two Astar 350 F2X helicopters ferry supplies to the Billiken Drill Camp 17 kilometres to the west on the Double Eagle Property of Noront Resources. The pilots also shuttle between drill rigs, including Noront's three rigs, working in splendid isolation in a flat landscape of snow and swamp and stunted trees. No one walks far here, except the claim stakers and the line cutters, and they too are scattered around the landscape and serviced by helicopter.

Neil Novak has abandoned his Toronto office for a few days in the James Bay Lowlands. After Noront struck promising nickel-copper massive sulphide in its first two drill holes late last August, a staking rush ensued. A small-scale claims map (issued by Intierra Mapping) shows a large horseshoe, open to the west, the different colours representing the junior miners active in the area known as McFauld's Lake. The lake itself, on the map, is a tiny clump on southeast side of the horseshoe that has been dubbed the Ring of Fire. The dominant colour on the ring is orange, representing the land position of Noront Resources Ltd.

Billiken Management is providing several services, including

Aerial view of Billiken Base Camp at McFauld's Lake. Looking east on March 26, 2008.

geological services. One geologist, Howard Lahti, is devoted to property of Spider Resources Inc., and two, Russell Crosby and Carl Archibald, are based in the Billiken Drill Camp, also known as the Noront Camp or the Eagle One Camp. Mike Kilbourne, project manager as well as professional geologist, oversees all of Billiken's activities for its clients at McFauld's Lake.

To begin the interview, Novak says, "From both companies, Spider and Noront, I'm here to get updated with the current drilling results. Both companies are drilling on their respective projects."

He flew in yesterday from Thunder Bay and is leaving tomorrow afternoon. The throbbing of a helicopter landing only 30 metres away threatens to drown out conversation. A helicopter takes off or lands every half hour or so.

Novak continues: "When I leave I do an update for the market by creating a press release for both companies."

Neil Novak of Noront Resources points to McFauld's Lake on the Intierra Mapping claims map. The writer assists. Jim Guillemette Photo.

Noront's most recent press release on McFauld's Lake Area, February 25, announced the discovery of a second nickel-copper massive sulphide occurrence, 2 kilometres southwest of the Eagle One deposit. It has been christened Eagle Two. Novak explains that the companies he represents like to have an executive actually visit the field to verify what is happening before they release new information. Noront is also drill testing several other high-priority geophysical targets.

Novak discloses his long association with this roadless area of Northern Ontario, dating back to the early '90's. After the De Beers Canada Inc. diamond discovery at its Victor Project between McFauld's Lake and the fly-in community of Attawapiskat on James Bay, Novak began an association with Spider Resources to explore for diamond-bearing kimberlites from 1992 to 1997. In 1995 he joined the board of Spider. The exploration history is complicated, but in 2002, Spider, in joint ventures with KWG Resources Inc. (KWG:TSX-V) and De Beers, drilled a geophysical anomaly near McFauld's Lake, encountering volcanogenic massive sulphide (VMS) containing copper and zinc. After the announcement in November, 2002, the area's first claim staking rush ensued. Novak had joined Noront's board in early 2002.

Then came Noront's Eagle One discovery last August, magmatic massive sulphide (MMS) containing nickel-copper and precious metals. It was a very exciting discovery, and sparked the second staking rush.

Novak says Noront's staking is ongoing as it infills and extends its holdings. Noront has entered into joint ventures with other juniors. "Noront would like to have every one of those properties here," he says. As he points to the map, his eyes gleam.

With regard to the nearest First Nation communities, Webequie and Marten Falls, many kilometers away, Novak says, "We've got very good rapport with both communities." Winter roads link both communities to the outside for a few months each year, and Novak would like to see the roads extended to the Double Eagle Project area. With regard to providing other services, "We try to get them involved in all aspects of exploration," he says.

Noront is in process of outlining a base and precious metal resource at Eagle One. Novak describes the splendid results on hole 27, announced on January 28. Over 35.6 metres, hole NOT-07-27 produced assays of 7.91% nickel, 3.45% copper, 1.66 grams per tonne (g/t) platinum, 12.97 g/t palladium, 3.87 g/t gold, and 9.27 g/t silver. With

Brian Newton, Mike Peplinski and Neil Novak stand by a key claim post on the north shore of McFauld's Lake, erected in the 2002 staking rush. Newton is co-owner of Billiken Management, Peplinski owns Greenstone Exploration (he planted the post), and Novak is Vice President/Exploration of Noront Resources. Jim Guillemette Photo.

such grades, Novak says, the rock would likely not require concentration on site before being shipped out.

Novak resumes, choosing his words carefully, "If there was a million tonnes of that grade of rock out there, and we don't have an official number – it could very well be much more than that – at that grade a million tonnes is worth in the order of $2.5 billion."

Asked what Noront hopes to find, Novak laughs, "Hundreds of millions of tonnes . . . I hope!" But he is realistic. "Something in the 8-to-10-million-tonne range, it would really make sense up here. At those grades that would be a phenomenal deposit, world class."

Examining the public record, one finds that in the fall of 2007, Noront's stock traded at record highs, closing at $6.66 on October 25. Earlier in the year closures were as low as 38 cents. Spider, at one point closing at 3 cents, closed at $0.18 on October 23.

Novak remarks on Noront's recent financing, for he feels the company is prepared. "The burn rate right now in Noront," he says, "is just under $2 million a month up here in our current program." That includes an airborne geophysical survey that has been running since the fall.

Looking to the future, Novak says, "The majors would like to come in here and make this a mining camp. That's their style." He adds: "We're listening to everybody, and we're talking to all the majors. And we didn't go to them, they came to us."

Novak puts on his other hat now. The President of Spider Resources is quick to point out the VMS occurrences and diamond kimberlites that Spider has identified, which are now more interesting. "If Noront has a mine," he says, " that means the logistics of the area have been overcome." Then, as for the development of Spider's properties, "The economics start making real sense."

At one point, John Harvey, P.Eng., Executive Vice President of Noront, enters the kitchen tent. He has been to McFauld's three times in six months. He explains that the Toronto office gets the logs of drill core online, but, he says, "We have to look at the real stuff once in a while." Then he volunteers: "We hope to announce Eagle Three next."

So, for Neil Novak, Noront Resources is one of the keys. He speaks of finding Eagle Three, then Eagle Four, then . . .

A geologist pulls him away, wants him to look at some interesting new core. The interview ends.

On March 26, 2008, the price of Noront stock closed at $5.39, and that of Spider at $0.12. Dreamers imagine that they have only one way to go, and that's up. Way up.

- Northern Ontario Business

MASSIVE SULPHIDES AT McFAULD'S LAKE*

Early on March 26, the plane banks east over McFauld's Lake, turns 180 degrees, and then aligns itself to approach the ice strip stretching along the north shore. Further east, the Attawapiskat River winds towards James Bay beneath its icy crown.

Banks of drill core racks stand out from the tents at McFauld's Lake base camp. A turboprop Caravan sits on the ice landing strip. Jim Guillemette Photo.

The Cessna Grand Caravan rolls up to clusters of fuel drums and lumber. Snowmobiles transport the supplies in sleds to the base camp just steps beyond.

The camp is run by Billiken Management Services, headquartered in Toronto. Billiken provides services to a number of mining exploration companies in the McFauld's Lake Area. In 2002, Spider Resources Inc. (SPQ:TSX.V), in joint venture with De Beers Canada Inc., was drilling on a magnetic anomaly, hoping to find another diamond-bearing kimberlite. Instead they found volcanogenic massive sulphides (VMS) with high concentrations of copper and zinc. In January, 2003, a staking rush began, bringing junior miners into the area. A number of exploration companies, including Noront Resources Ltd. (NOT:TSX.V), acquired land positions in the greenstone belt.

The hub of the McFauld's Lake base camp is the kitchen tent. Unlike the office tent next door, the dining area hums with activity. Two cooks, in their tiny preparation area, provide three square meals a day, feeding 40 people. At all hours, people enter to grab a coffee and a snack. Workers debrief one another. Managers conduct business.

Mike Kilbourne, Billiken's project manager, and project geologist for Noront, arranges for the writer and the photographer, Jim Guillemette, to fly to Noront's Eagle One Camp, also known as the Drill Camp. Every half hour or so, an Astar 350 FX2 helicopter takes off or lands only 30 metres away. Walking or sledding for any great distance is out of the question in this roadless wilderness.

The bright red helicopter, piloted by Sam Lappage of Cochrane, Ontario, lifts off the wooden landing pad. The camp, with its dozen and a half tents, falls away. A prominent feature are banks of core racks, where signs warn off trespassers. The writer has not been invited to the core shack. Two Noront executives are in camp, examining drill core, preparing to making public the company's latest findings.

The chopper's passengers include Russell Crosby and Carl Archibald. Crosby, P.Geo., and Archibald, junior geologist, are returning to their base at the Eagle One Camp, 17 kilometres to the west. In August of 2007, Noront Resources was drilling a magnetic anomaly on its Double Eagle Property when it discovered a different kind of deposit, magmatic massive sulphides (MMS). The core contained spectacular concentrations of nickel (unlike VMS). A second staking rush occurred. Companies staked thousands of square kilometers in the area now known as the Ring of Fire. On a claims map the "ring" resembles a crescent, open to the west.

In late January, 2008, Noront released the assay results of hole NOT-07-27. Over 35.6 metres, the hole produced assays of 7.91% nickel, 3.45% copper, 1.66 grams per tonne (g/t) platinum, 12.97 g/t palladium, 3.87 g/t gold, and 9.27 g/t silver. In late February, Noront announced the discovery of a second nickel-copper massive sulphide occurrence, 2 kilometres southwest of the Eagle One deposit. It had been christened Eagle Two. By press time, Noront had still not released assay results due to heavy demands on the independent lab.

On March 26, the pilot flies low enough for passengers to spot snowshoe tracks in the gridwork of lines cut out of the trees. Vast empty, snowy stretches of bog alternate with patches of forest. Lappage detours to a diamond drill rig that is being relocated. Another red chopper, piloted by Rejean Dulong of La Sarre, Quebec, is lifting a section of the dismantled rig by means of a very long cable, and transporting it a few hundred metres for re-assembly on a new site.

A few minutes later, Lappage sets the helicopter down gingerly in a tiny clearing of a black spruce stand. The trees are only twice as high as the chopper blades off the ground; the snow is probably knee deep. The geologists pad down a snowmobile track to an operating rig. This is named the Johnny Cash, Crosby says. The one we saw being relocated was the June Carter Cash, both names paying tribute to the late singer, as does the moniker "Ring of Fire".

The Johnny Cash rests on two cross-hatched tiers of heavy square timbers in turn sitting on tree trunks. If the rig were sitting on the frozen ground, the heat and vibration generated by the rig would cause it to sink. Crosby makes a quick examination of the drill core, which was pulled from a depth of 189 metres. The hole is NOT-08-1G20, and the drill is slanted at an angle of -90 degrees, that is, at the vertical.

A few minutes more flight time brings the chopper to the Eagle One Camp, nestled in a mixed forest dominated by black spruce. The camp spreads over a rocky knoll. The terrain now resembles Precambrian rock and boreal forest, for the geographical boundary of the Lowlands runs north-south in this area.

On the right, geologists Carl Archibald and Russell Crosby discuss the core just retrieved. Inside the drill rig, the racket impedes conversation. On the left, Jim Guillemette, photographer.

Inside the drill rig, Russell Crosby, P.Geo., examines the latest core pulled from a depth of 189 metres. Jim Guillemette Photo.

Billiken Management built the camp last fall. It looks newer and neater. Tents predominate, but some cabins are constructed of wood. The geologists give a tour of their domain - three double-wall tents stretched on wooden frames. One is a bunk cabin, and two are core shacks - one for logging, and one for cutting. Sitting on a wooden floor, a stove heats each tent, fueled by diesel oil.

A helicopter brings anywhere from 10 to 40 core boxes a day suspended in a sling. Geologists log the core, that is, describe its characteristics, centimeter by centimeter. A special saw cuts some core lengthwise. Half the split cores are sent out for assay, the remainder are stored.

Crosby says that the overburden at a drilling site varies from 5 to 30 metres in depth. Then the drill sometimes encounters a layer of limestone, up to 10 metres thick, and then possibly a layer of sandstone of similar thicknesses. Limestone and sandstone are Paleozoic rocks. The geologists are seeking Precambrian rocks, specifically peridotite, the host rock for MMS.

The pilot Dulong flies the writer and photographer back to base camp. He remarks, "It's the place not to be in summer." He is serving his seventeenth tour in the Lowlands, and this summer he will likely pass up the boot-sucking bogs, hordes of mosquitoes, and landing pads in the bush, constructed of logs, that swiftly tilt to dangerous angles.

At base camp, the office (left) differs from the bunk tents only on the inside, where there is a desk to accompany the project manager's bunk.

Earlier in the camp kitchen, Neil Novak, Vice President/Exploration for Noront Resources, and President of Spider Resources, had spoken at length about Noront's prospects. Referring to hole 27, described above, he stated that 8 to 10 million tonnes with such grades would justify a mine, making the deposit "world class". The ore would likely not require concentration on site before being shipped out.

John Harvey, Executive Director of Noront, commented that the MMS deposit had resemblances to the deposits at Sudbury and Voisey's Bay. Like Novak, he was highly optimistic: "We have tens and tens of anomalies to be [drill] tested," he said.

In April, the Acting Provincial Mining Recorder in Sudbury commented on the staking rush last fall. "The large influx did put a lot of pressure on our resources," said Stephen DeVos. In 2007, Ontario recorded 122,808 claim units. About a sixth of that number related to the Ring of Fire, covering about 3,600 square miles.

A piece of the Ring of Fire is the responsibility of Thunder Bay North District of the Ministry of Northern Development and Mines. Mark Smyk, Regional Resident Geologist, said, "We kind of have the horns of the crescent," referring to the western edge of the area. He calculated there were about two dozen companies exploring in the area, many of them in joint ventures. Will anything be happening soon? "[The companies] are waiting to see what happens with Noront . . . They're feeling pretty bullish about it."

In the office tent at McFauld's Lake, Mike Kilbourne, project manager, takes a few minutes to explain his job. He oversees procedures to ensure core samples arrive at the assay lab through an unbroken chain of custody. He spots (locates) drill hole collars for the rigs. He manages seven or eight different contractors, such as kitchen staff, geologists, geophysical survey crews, and helicopter personnel. He draws up budgets and manages accounts.

For the third or fourth time that day, a Caravan arrives with cargo, and the reporters catch a ride out. The plane lifts off just before seven. It is still light. When it lands at Nakina airport 183 miles away, an hour and ten minutes later, the sun has dropped below the horizon.

One imagines the lights gleaming from scattered camps and drill rigs throughout the Ring of Fire.

*- *The Chronicle-Journal, under the heading "Empire in training : Mining company goes to work in northwest"*

RING OF FIRE SMOKIN' : GEDC Annual General Meeting

Many people from Greenstone region are working in the Ring of Fire, flying into the James Bay Lowlands from Nakina Airport. Two speakers at the May 6 meeting dealt with the exploration boom.

Greg Stott, Geoscientist with the Ontario Geological Survey, spoke of the exploration opportunities in the Lowlands. Most of the presentation appealed mainly to the geologically literate in attendance. The Ring of Fire, also known as the McFauld's Lake Area, follows a greenstone belt shaped like an enormous crescent moon open to the west.

In summing up, Stott stated that the region in and near the Lowlands is still a geological frontier, and that Precambrian geology underlies the cap rock of the Lowlands. Targets in the area include the following: copper, nickel, platinum, palladium, gold, and in kimberlite pipes along several possible corridors, diamonds.

Mike Kilbourne is Project Manager for Billiken Management Services Inc. working out of McFauld's Lake base camp. Billiken is an exploration management consulting company based in Toronto. One of its most important clients in the Ring of Fire is Noront Resources Ltd. Its main property is the Double Eagle, just west of McFauld's Lake.

In August 2007, Noront's drills intersected massive copper-nickel sulphides at Eagle One, and the company is in process of estimating the resource to make it compliant with NI 43-101. Within five kilometers of both Eagle One and Eagle Two, Noront has multiple high-priority targets to drill. The Ring of Fire itself, in which up to two dozen companies are exploring, is 150 kilometres long.

The community of Webequie is 59 km west; Marten Falls, 115 km south; the De Beers' Victor diamond mine, 170 km east. All the foregoing have a winter road. Aircraft provide year-round access to mining properties, with the Greenstone region being centrally located for that purpose.

Kilbourne concluded by stating that the Greenstone region can supply products, services, and skilled labour to meet mining exploration requirements.

- Times Star

Leading the Charge
RING OF FIRE
James Bay Lowlands, Ontario
Noront continues to be the dominant land holder in the Ring of Fire with high grade nickel, copper, PGM's and chromite discoveries.
Suite 1000, 15 Toronto Street
Toronto, ON, Canada M5C 2E3
Tel: 416 367 1444
Investor Relations: 416-238-7226
www.norontresources.com
NOT: TSX-V
NORONT
RESOURCES LTD.

SPIDER resources

Spider is the Pioneer Exploration Company in the James Bay Lowlands, exploring since 1993. Responsible for the discovery of Three Kimberlites (MF1, MF2 and Good Friday) within a few kilometres of Ontario's first diamond mine, as well as five hidden kimberlites (Kyle 1 – 5) located 100 kilometres to west. Regional diamond exploration in 1996 and 1997 lead to discovery of McFauld's Lake VMS deposits (1 and 3) in 2002, as well as 8 other VMS occurrences in the McFaulds Ring of Fire area. Geophysical surveying and other exploration efforts as conducted by Spider formed the exploration data set that eventually lead to the discovery of the Eagle One Magmatic Massive Sulphide Deposit.

Please visit us at the PDAC, March 1 to 4, 2009 – Booth 3127!

For information, contact Neil Novak
Tel: 416.815.8666
Fax: 416.815.1355

470 – 44 Victoria Street, Toronto, Ontario M5C 1Y2

www.spiderresources.com

Photos by J. Guillemette

Chapter 6 : Year 2008

Société du developpement économique de
GREENSTONE
Economic Development Corporation

Activities of Greenstone Economic Development Corporation

Muskeg Tours 2009

Presenters at Mining Procurement Workshop II on October 22nd

MINING PROCUREMENT WORKSHOP

Mining Exploration Overview

The residents of Greenstone will look back on these days as historic milestones. The region east and north of Lake Nipigon has never seen this level of mining exploration since the 1930's.

On Thursday, March 20, Greenstone Economic Development Corporation (GEDC) sponsored a mining procurement workshop at the Geraldton Curling Club. About 90 people, the majority of them local business people, with a sprinkling of aspiring entrepreneurs, turned out.

The workshop offered an opportunity to network with mining exploration companies and contractors, to learn how to bid on product and service delivery, and to learn how to promote local products and services useful to the mining sector.

Dina Quenneville, executive director of GEDC, and Kathy Durocher, president of the board, welcomed presenters and participants. The Ministry of Northern Development and Mines (MNDM) delivered the first presentation.

Gerry White, MNDM district geologist for Thunder Bay North, embracing the area east, north, and northwest of Lake Nipigon, spoke on mining exploration activity.

White presented an overview of activity, pointing out that over $500 million has been spent recently, much of it in Northwestern Ontario (NWO). There have been record levels of claim staking in the pursuit of gold, base metals, uranium, and diamonds. There are currently in excess of 300 active explorations in NWO alone.

The Thunder Bay North district stands out, the Beardmore-Geraldton gold belt in particular. White projected that in the year 2008, $20 million will be spent in the Greenstone region.

Kodiak Exploration Limited is the biggest player in the Greenstone region. The company will soon enter its third year of exploration of the Hercules Project. Many other companies, attracted by its successes, are involved in staking and exploring.

"All the prospectors in Beardmore," said White, "have optioned every bit of ground they've staked. It's unprecedented in that respect."

White cautioned that exploration does not inevitably lead to a mine. There are currently five producing mines in NWO. Statistics tell us that only one in 10,000 occurrences (promising mineralized ground) yields a mine. But, said White, "All it takes is one mine to make a difference."

White led the participants through the process of mining exploration, beginning with staking. In the front line are the prospectors. Unlike formally trained geologists, who tend to target promising ground, prospectors look at everything. "Gold can be found in any rock type," said White, "as opposed to [occurrences of] copper-lead-zinc."

Gerry White, district geologist for MNDM, points out the red blocks that dominate the map "Beardmore/Geraldton Mining and Exploration – February 2008" issued by Intierra Mapping. They indicate the land position of Kodiak Exploration. Jack Bolen, project geologist for Sage Gold, puts his finger on his company's Onaman property.

The prospector with a shovel and a grub hoe gives way to more specialized tradesmen who do geological mapping and geophysical surveys.

Mining exploration companies in Greenstone region are involved in activities such as stripping, trenching, sampling, drilling, linecutting, mapping, and surveying. These activities suggest career possibilities for young people, such as geologist, geographic information system (G.I.S.) specialist, and helicopter pilot.

Exploration companies lay out money locally for equipment, accommodation, food, fuel, contractors, employees, and certain services such as expediting and transportation.

White referred to several companies active in this area. Kodiak Exploration received appropriate attention. "The key property that sort of got everything going is the Hercules Project in Elmhirst Township," said White. In 2007, Kodiak spent $1.1 million there. Its drilling program intercepted many spectacular gold values, such as 32.96 grams per tonne (g/t) over 11.6 metres (m), and 358.56 g/t over 3.6 m. "That's pretty impressive in this business," he said.

The Geraldton Project of Premier Gold Mines Limited and Roxmark Mines Limited, said White, is proving the old adage: "The best place to look for new gold is where you found the old gold."

White referenced several other companies and prospects, most of them dealing with gold values:

O'Sullivan Lake Project of Superior Canadian Resources;
Junior Lake Project (nickel; copper) of Landore Resources Limited;
Kilometre 61 Property (molybdenum; copper; silver) of Linear Metals Corporation;
Marshall Lake Project (copper, zinc; silver; gold) of East-West Resource Corporation and Eyeconomy Holdings PLC;
Jacobus Property (nickel; copper) of Sage Gold Inc.;
Onaman Property (copper; lead; zinc; gold; silver) of Sage Gold Inc.;
Northern Empire Mine Property of Roxmark Mines Limited;
Nortoba-Tyson Project (molybdenum) of Roxmark Mines Limited;
Brookbank Property of Ontex Resources Ltd.;
Mud Lake Property of Alto Ventures Ltd. and Wescan Goldfields Inc.;
Knucklethumb Property (Norah Lake Property) of Spruce Ridge Resources Ltd.

White concluded his presentation by observing, "There are other companies working in this area that have land positions [claims]."

White did not speak about the exciting activity in the McFauld's Lake area north of the Albany River, on the Attawapiskat River. There has been a staking rush in pursuit of copper-nickel deposits and of diamond pipes. The airport in Nakina is bustling nowadays.

The workshop started at 9:30 a.m. and ran until about 4:00 p.m. There were generous breaks for networking activities and for visits to exhibitor booths and for food and beverages. The level of excitement in the room remained high throughout, and in the end, participants exhibited a reluctance to leave.

Junior Miners Speak Out

On March 20, several junior mining companies active in the region described their programs. Greenstone Economic Development Corporation sponsored the workshop in Geraldton.

Sage Gold Inc.

Jack Bolen, project geologist, said that Sage Gold had two backhoes operating, and two drill rigs on its Onaman property exploring for base and precious metals. "We have defined," he said, "from 100,000 to 200,000 tonnes in the last month, but we are still only 10% of the way to being economic." That is, to finding a mine. He stated that Sage was at minimum two years away from that happy state, and at minimum five years away from a producing mine. He stressed the word "minimum".

Sage hired a line-cutting crew of 12 from outside the region because there was no local contractor available. "I'm dealing with contractors, not individuals," he said. He simply could not personally supervise them in the field.

As for business people, he suggested that they should not be tying up their capital in anticipation of quick results from exploration. "But," he said, "you must plan so you can move on short notice." If a mine were developed, there would be many skills required, including occupations such as mechanic, carpenter, electrician, and bookkeeper – the same jobs now essential to the forest industry.

Bolen went on to say that if a mine were developed, Thunder Bay would still be very important for supplies and services, but the company would have a satellite office in Greenstone.

Northstar Drilling Ltd.

Facilitator Dina Quenneville introduced Barbara Courte as the only female in North America who owned a diamond drilling company. Courte said that Northstar Drilling had five drill rigs and 35 employees.

In the last two years, her company had landed contracts with junior miners active in the Greenstone region, including Ontex Resources, Alto Ventures, East West Resource, and Linear Metals. As she spoke, she had two rigs operating on Sage property and two on Kodiak Exploration. The fifth rig was in Shebandowan.

She explained that she could find no local drillers to employ, but was open to hiring driller's helpers. She stressed that a helper must have a strong work ethic to work a 12-hour shift for weeks at a time. "It is the hardest job you can do," she said, "and in order to be a driller, you have to be a good helper."

North Star rented cabins, and could employ local people for such jobs as cooks, mechanics, and welders.

Roxmark Mines Ltd.

Louis Bernard, Senior Vice President/Operations, briefly outlined his company's history and pointed to its custom mill in Beardmore, the only one in the region.

Glenn Hart, geologist, explained that their molybdenum property, Nortoba-Tyson, had also produced assays for high-grade gold and for some copper.

Roxmark was exploring the old Northern Empire gold mine, and had optioned property to the west. Hart remarked, "Everything's about tonnage and grade in the mining industry."

On the old mine property, the Power Zone was mined out, but the Contact Zone, 10 metres to the north, had never been touched. A drilling program last year, exploring to the east, found economic grades in the Power Zone near surface. Oldtime miners "followed the veins", he said, and did little exploration or planning for the long term.

Roxmark was not hiring, but welcomed submissions of workers' names and skills. "Forestry workers make good workers, even in the mining industry."

The company was planning to buy its own drill rig, and would require people to operate and maintain it.

Referring to the old Northern Empire, Hart said that the old shafts and underground workings needed considerable upgrading to meet current standards. "If we could identify the resources underground," he said, "we could probably have a mine within a year."

Roxmark also controlled the properties of the old Sand River and old Leitch mines.

Premier Gold Mines Ltd.

Melodie Stone, geologist, spoke on the company's Geraldton Project. Over the winter they had had a drilling program on the old Magnet gold mine property, but were now focusing on the old Little Long Lac mine property, "with most exciting results". At that date, Premier Gold had spent $1.7 million.

Premier had discovered two new veins at Little Long Lac, and had two rigs operating there.

As for needs of Premier that have already been satisfied by local businesses, Stone presented a detailed list in PowerPoint. The list included safety gear, office supplies, expediting, heavy equipment, core boxes, and fuel.

She also emphasized the efforts of the company to maintain good relations with the municipality and with residents near the drilling sites.

Kodiak Exploration Ltd.

Martin (Marty) Laforest, operations manager, referred to the company's escalating spending in this region. His talk focused on issues surrounding the exploration phase (trying to find an economic resource) and the development phase (developing a mine after an ore body is established).

In exploration, a junior miner selects an area or region, screens targets, and then acquires a land position (control over property). Then come the steps, such as drilling, to discover economic deposits.

In development, issues include the ore depth (a technical issue), regulation and licencing (political), community involvement (social), and price fluctuations (economic).

Laforest outlined the undetermined risks of the future: an increase in the depth of the ore body, an increase in social and political risks, a decrease in demand for the product (e.g., gold), and decreases in budgets of the major mining companies.

Linear Metals Corp.

Matt Rees, chief geologist, described the company's Kilometre 61 Project, east of Armstrong. He predicted a growing demand for molybdenum. "They're coming up with new uses all the time," he said, and high prices are driving the industry.

Linear Metals found the resource in 2002, north of Lake Nipigon. "Logistically," he said, "we're a bit more tied to Thunder Bay." The company anticipated an open pit operation.

In June their drilling program would wind up, and stripping and trenching would commence.

His advice to the participants: "Get out and introduce yourselves. We certainly don't know all the suppliers out there." He mentioned meeting, at the recent Toronto conference of Prospectors and Developers Association of Canada, a local supplier of core boxes, Peter Kathmann.

Other Presenters

Daniel Nolet, foreman, Chibougamau Diamond Drilling, made himself available for one-on-one's. Mike Peplinski, camp foreman for Billiken Management Services, was introduced to the meeting when he was called away on business. Billiken supplied services to several junior miners in the McFauld's Lake Area, north of Nakina in the James Bay Lowlands.

Further information on the companies can be found on their websites.

Non-Miners Offer Advice

On Thursday, March 20, Greenstone Economic Development Corporation sponsored a mining procurement workshop at the Geraldton Curling Club. Three speakers not directly involved in the mining industry offered some advice.

The Administrator of Greenstone Community Portal explained the feature called Greenstone Marketplace. Jay Lucas urged all businesses to register, free of charge. Each participant may showcase his or her products and services, and provide contact information including a link to one's business website.

Lucas, an employee of Geraldton Community Forest Inc., emphasized that the Portal is truly a community site, open to every business and community group. "We're getting thousands of hits a day on our site," he said. World-wide Internet searches can lead a prospective customer to a Greenstone resident. A business website, he said, is a popular way of doing business nowadays.

Any business can apply for the free Greenstone Community Portal Business Toolkit up until July 31.

Blair Smith, of BDO Dunwoody, weighed in with some pithy financial advice. And Kim Tarantula, agent for CAN Canada, promoted her company's picture telephone as a way for mining people to keep in touch with home or business. She said that the device requires access to a high speed Internet service.

- Times Star

TWENTIETH ANNIVERSARY AGM

Greenstone Economic Development Corporation recently celebrated its twentieth anniversary of supporting and diversifying the region's economy. The annual general meeting of the non-profit organization took place at the French Club in Geraldton on Tuesday, May 6.

Over sixty members and guests attended, not counting the thirteen GEDC members of the Board of Directors and staff. In her report, President Kathy Durocher explained that GEDC evolved from the BGLN Economic Development Centre, established in 1988. In 1995, GEDC adopted its current name.

In the past year, said Durocher, GEDC completed key initiatives of the Greenstone Recovery Project, comprising the Recruitment and Retention Team Project, the Destination Marketing Committee, the Canada Yew Cultivation Study, the Kinghorn Rail Study, and the Wood Densification Study.

Another GEDC initiative is the Data Modelling Project, a development tool for the mining industry. The project aims to provide 3-dimensional modeling of former operating mines. Partners in the project include the local mining sector, the Geraldton Chamber of Commerce, and the Mines and Minerals Division in Thunder Bay.

GEDC staff members are Dina Quenneville, Executive Director; Sylvie Goulet, Business Development Officer; Jamie Armstrong, Project Analyst; Shelly Stanert, Administrative Assistant; and Megan Wilson, Ontario Self-Employment Benefit Business Advisor.

The Facilitator, Dina Quenneville, confirmed the following continuing positions on the Executive: Kathy Durocher, President; Lynn Gosselin, Vice President; James Slattery, Secretary/Treasurer; Michael Power, Past President; and the following Directors, James Slattery (IT Sector), Ross Houghton (Forestry Sector), Glenn Hart (Mining Sector), Mary-Jane Dunn (Banking).

There being no nominations for Director to represent the Insurance Sector, Kathy Durocher agreed to represent that sector. In the election for Director to represent the Entrepreneur Sector, the meeting confirmed Lynn Gosselin.

Kathy Durocher, President of GEDC, delivered her annual report on May 6. GCHS Knightline Photo.

Vice President Gosselin introduced the past year's recipients of the Local Initiative Contributions (funding):

Geraldton Walleye Classic and Kenogamisis Fish & Game Sno Club (John & Diane Magee)
Geraldton Area Natural Resources Advisory Committee (Gerard Dufour)
Caramat Community Club Inc. (no representative)
Geraldton Concert Series (Frank Dzijacky)
Longlac Historical Society (n.r.)
Greenstone Fall Fair Committee (Karen MacDougall)
Paro Centre for Women's Enterprise (Lori Beaulieu)
Kenogamisis Recreation Corporation (Joe Audia)
Aroland First Nation (n.r.)
Ecole Secondaire Chateau-Jeunesse (Francois Sarrazin)
Longlac Ski Club (Terry Jankowski)
Geraldton Nordic Ski Club (Ken House)
Longlac Senior Citizen Club (Madeline Owca)

A number of speakers and presentations rounded out the meeting. Times Star will present summaries in upcoming articles.

Three students from Geraldton Composite High School filmed the proceedings for Knightline, the local community television program. Nicki Dumont, Leena Thibodeau, and Patrick Levesque also conducted short interviews. They would be sending the footage to Thunder Bay Television for editing and for airing on its weekly Northwest Newsweek program on Sunday evenings.

The winner of the draw for a Klaus Rossler print of Lake Superior was Martin Laforest.

Facilitator Quenneville remarked at the close of the meeting that Greenstone's mining sector is very strong. She pointed to the evening's presentations, and to published remarks of such personages as Dr. Robert Rosehart, Economic Facilitator for NWO; Chris Hodgson, President of Ontario Mining Association; and Jason Goulen, Director, Metals Economics Group.

Goulen had delivered a report, "World Exploration Trends", for the PDAC International Convention 2008 in early March in Toronto. GEDC participated in that Convention, which saw some 22,000 people enjoying the exhibits.

GREENSTONE INITIATIVES

Members of Greenstone Economic Development Corporation learned more about the organization's regional initiatives at their AGM on May 6.

Megan Wilson, Business Advisor for Ontario Self-Employment Benefits, outlined her program. Clients receive support over a period of 48 weeks for the launching and operation of their businesses. Clients must first submit a comprehensive business plan. In 2007 five clients successfully started businesses: Darrel Ward (ECT), Chris Walterson (Olden Grey Reno), Luisa Wilson (A Design in Time), Lori Beaulieu (Star Consulting), and Barbara Gregorin (Barking Heaven).

The local owner of Northern Earth GIS Services presented GEDC's Green Energy Initiative. Corby Richmond discussed solar air heating, solar water heating, geothermal energy, solar electricity, and wind energy. He explained that energy efficient practices and renewable energy are within reach of Greenstone residents.

Jamie Armstong, Project Analyst, explained that GEDC's mining project entails working closely with local mining exploration companies. Greenstone Mineral Development Initiative has recently received funding from FedNor . The first phase has focused on Summers and Eva Townships in the Beardmore area. Armstrong presented two 3D models of former operating mines, the Leitch and the Northern Empire, both properties now owned by Roxmark Mines Ltd. A steering committee will render guidance for the second phase.

Mayor Michael Power introduced a miner who had experience in the old mines of Little Long Lac, Macleod-Cockshutt, and Mosher. Siegfried "Siggy" Neumuller approached the podium to receive a framed photograph of the Macleod-Cockshutt headframe. After the standing ovation, Neumuller described how he started underground in 1951 and engaged in such work as driller and hoist man, until he was laid off in 1967 as the last operating mine was phasing out operations.

At this point in the meeting, Dina Quenneville, Executive Director, played the Mining Resurgence video aired by Thunder Bay Television on Northwest Newsweek. The video reported on the Mines and Minerals Symposium of April 8 and 9.

DISTRICT MINING REPORTS

Several "miners" working in the Beardmore-Geraldton Greenstone Belt made presentations to the AGM of Greenstone Economic Development Corporation on May 6.

Gerry White, District Geologist for the Ministry of Northern Development and Mines, spoke about the unprecedented amount of money being spent by more than three dozen companies in the region.

The Ontario Geological Survey is monitoring 76 active projects in the province, with gold (Au) accounting for 54% of the activity; copper (Cu)-nickel (Ni)-platinum group elements (PGE), 21%; Cu-zinc (Zn)-silver (Ag), 13%; diamonds, 3%; and the remainder divided among uranium and industrial minerals.

White's slide presentation presented an overview of activity in Thunder Bay North District, as follows:

Musselwhite Mine (Goldcorp Inc.) - Au

Numerous junior mining exploration firms in the "Ring of Fire" of the James Bay Lowlands, looking for Cu, Ni, Zn, PGE, and diamonds.

Melchett Lake Property (Stratabound Minerals Corp.) – Zn, Ag.

Hercules Project (Kodiak Exploration Ltd.) – Au

Geraldton Project (Premier Gold Mines Ltd.) – Au

O'Sullivan Lake Project (Superior Canadian Resources Ltd.) – Au

Junior Lake Project (Landore Resources Ltd.) – Cu, Ni

Kilometre 61 Property (Linear Metals Corp.) – molybdenum (Mo), Cu, Ag

Marshall Lake Project (East West Resource Corp.) – Cu, Zn, Ag, Au

Jacobus Property (Sage Gold Inc.) – Cu, Ni

Onaman Base Metal Property – Cu, lead (Pb), Zn, Au, Ag

Northern Empire Mine Property (Roxmark Mines Ltd.) – Au

Nortoba-Tyson Project (Roxmark Mines Ltd.) – Mo

Brookbank Gold Property (Ontex Resources Ltd.) – Au

Mud Lake Property (Alto Ventures Ltd. / Wescan Goldfields Inc.) – Au

Knucklethumb and Nora Lake (Spruce Ridge Resources Ltd.) – Au

Orphan Mine Property (Mantis Mineral Corp.) – Au

The presentation was prepared by White and Mark Smyk, Regional Resident Geologist.

The owner of North Star Drilling and of Cobra Drilling spoke about their partnership with Confederation College. Barbara Courte said that eleven students had just finished a driller helper program and were starting their first 12-hour shifts. Another group of students were just starting the program.

Larry Hillesland, Project Manager for Kodiak Exploration Limited, expressed the opinion that the Tashota-Onaman Greenstone Belt and the Beardmore-Geraldton Greenstone Belt have been underexplored. The price of gold is rising, he said, because the supply is decreasing.

Hillesland gave an overview of the Hercules Project, located north of Jellicoe up the Kinghorn Road. Kodiak is currently engaged in a 60,000-metre drilling program. The Project, according to aerial geophysical surveys, still has 30 kilometres of untested structures. The Golden Mile Zone alone has another 2 kilometres of unexplored structure. Surface exploration of these targets began in April.

Hillesland stated that Kodiak has control of 300 kilometres of major mineralized structures in the district. The structures extend from Beardmore-Lake Nipigon to Geraldton and further east. Many project-scale zones are ready for drilling. Three drill rigs are now operating on the Golden Mile and Marino Zones.

The geologist for Premier Gold Mines Ltd. gave an overview of his company's operations. Matt Long said that Premier has its flagship operation in Red Lake District. He spoke about its interests in the Musselwhite District and in Mexico. Premier also considers the Beardmore-Geraldton camp to be underexplored, hence its Geraldton Project. Recently it announced the potential of the Kailey Zone for an open pit operation on the old Little Long Lac mine property. Premier has a joint venture with Roxmark Mines Ltd.

Monir Younan, President of Roxmark Mines Ltd., stated that their main focus is the Beardmore area. The company's founder, Dr. Stan Malouf, passed away this year at age 94. Malouf was one of the first to recognize the unexplored potential of the Beardmore-Geraldton gold camp. He began exploring the old Magnet mine and then expanded afield to Beardmore.

In the Beardmore area, Roxmark is exploring for gold on the property of the past producers Northern Empire, Sand River, and Leitch. Roxmark has a 2008 drilling program in place. It also has a gold as well as a molybdenum prospect in Nortoba-Tyson.

In the Geraldton area, Roxmark has a joint venture with Premier Gold Mines Ltd. to explore past producers Little Long Lac and Magnet.

The Greenstone region also has an interest beyond its borders, in the James Bay Lowlands, where exploration is proceeding apace for minerals such as copper, nickel, and diamonds.

- Times Star

Jamie Armstrong, Project Analyst, has produced a 2D picture of a 3D representation of the workings of the old Leitch mine.

Geraldton has changed a bit since August of 1936

Stake a claim where the Past meets the Future

http://mining.gedc.ca

The Geraldton-Beardmore gold camp ranked among the top five producing areas in Canada, generating 4.1 million ounces of gold from 18.7 million tons of ore for an average recovered grade of 0.30 oz.Au/ton. The first production phase ended primarily because of a $35 fixed gold price and issues related to fragmented land holdings at the time.

Past Producers (1934-1968)	Tons Milled	Average Grade of Gold (oz/ton)	Gold Produced (oz)
Roxmark holdings			
1. Leitch Mine	920,745	0.92	847,690
2. Sand River Mine	157,870	0.32	50,065
3. Northern Empire Mine	425,866	0.35	149,493
4. Magnet Mine	359,912	0.42	152,089
5. Little Long Lac	1,780,516	0.34	605,499
6. Bankfield Mine	231,009	0.29	66,417
Total Roxmark holdings	**3,875,918**	**0.48**	**1,871,253**
7. Hard Rock	1,458,375	0.18	269,081
8. MacLeod-Cockshutt	10,337,229	0.14	1,475,728
9. Mosher Long Lac	2,710,657	0.12	330,265
10. Talmora-Long Lac	6,634	0.21	1,417
11. Jellico Mine	10,620	0.40	4,238
12. Tombill Mine	190,622	0.36	69,120
13. Dikdik – Orphan - Mantis	3,525	0.70	2,460
14. Sturgeon River Mine	145,123	0.51	73,438
Total Beardmore-Geraldton	**18,738,703**	**0.30**	**4,097,000**

Production Statistics courtesy of Roxmark Mines Ltd

Photograph: Northern Empire Mine, 1939 (Charles Dobie)

How '*Stake a claim where the Past meets the Future*' was born

In 2006, the GEDC Board of Directors tasked the staff with the mission of finding out how GEDC could assist with the Mining Sector revival as it continued to be known as *the forgotten gold camp*. Staff met and visited Kirkland Lake Gold Inc's Chief Geologist Mike Sutton to discover the data compilation activity compiled under his care which resulted in the re-opening of the Macassa Property. The resultant effect was the amalgamation of several mine properties under the ownership of Kirkland Lake Gold Inc. Mike Sutton provided GEDC staff with the necessary details and advice on how to begin the process of data compilation with the goal of 3D modeling. Staff returned with a proposal outlining the necessary steps in beginning the data compilation work. In February 2006, the Board of Directors agreed to support the project and provided the staff with the necessary resources to begin.

The *Stake a claim where the Past meets the Future* slogan was designed by the GEDC staff in an effort to communicate the vision of the sectoral project. The communities in the Municipality of Greenstone once thrived on gold mining. Many people believe the past will repeat itself and the area will once again be a miner's haven. GEDC is aiming to digitally compile all the mining sites and drill hole data in the Greenstone area and as such are working with exploration and mining companies of today and tomorrow to assist with 3D modeling. With the help of Datamine Studio, our objective is to provide the region with a 3D database which coincides with the Northern Development and Mines regional library, which is currently housed at GEDC in Geraldton. The Datamine Studio software was selected as it represented a focused investment in geological modeling and resource / reserve management and it also significantly expands our capability into new and exciting areas. The project continues to encourage new exploration in existing and new potential exploration sites. In fact, throughout the region, GEDC fosters and encourages the start-up and operation of successful enterprises in all economic sectors. It is GEDC's belief that the once-booming Greenstone belt will again thrive with the help of the current exploration activity and the many on-going projects in the area.

Inception Project Partners

Upon inception of the project, GEDC's project partners included the Geraldton Chamber of Commerce and Roxmark Mines Ltd. The key to the success of the project was established early on, data. Massive amounts of data were required. Without the data the project would fail. Through the direction of Monir Younan, President of Roxmark Mines Ltd. GEDC received the Leitch Mine property raw data in the form of paper drill logs and maps. Staff training began in May 2006 with Datamine Canada staff. It was at that time that the dissection of the Leitch Mine property data began. A full out data compilation and modeling process for the Leitch property commenced.

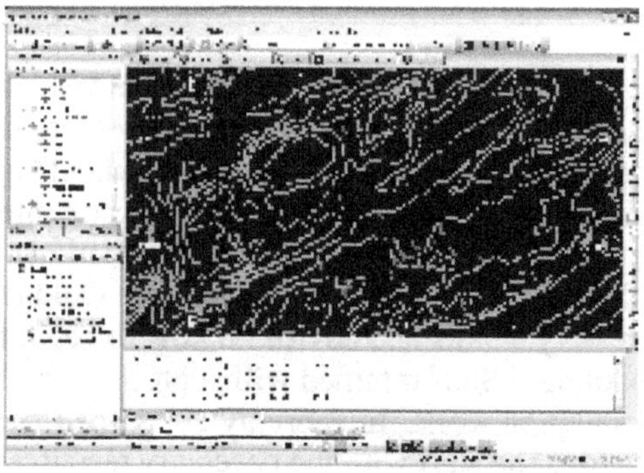

Image created using Datamine Studio 3

Datamine Studio provides a fully comprehensive and interactive three dimensional system for creating, modifying, merging, intersecting, displaying and evaluating both solid models and Digital Terrain Models (DTMs). The application uses a combination of modeling features to accurately represent geological structures and grade variations within the target area. After completion of the Leitch Mine property 3D model, the Northern Empire Mine 3D model was created.

Leitch Mine 3D Model (2007)

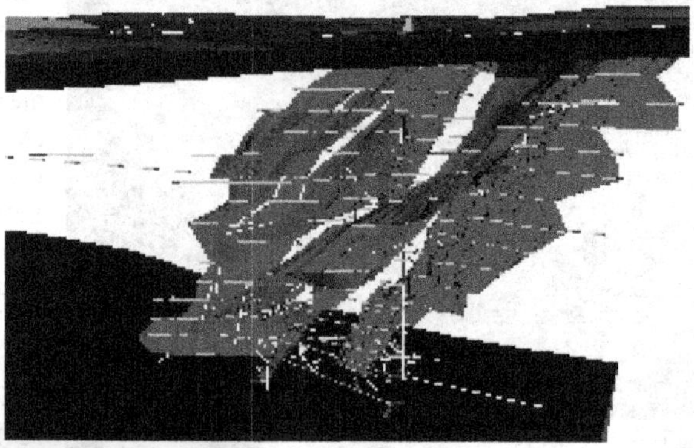

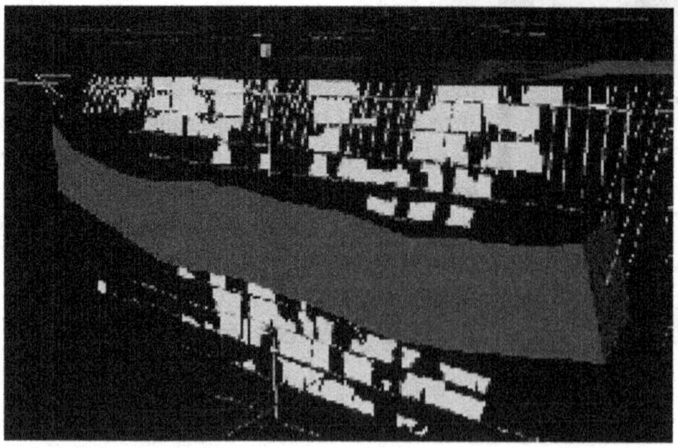

Northern Empire Mine 3D Model (2008)

GEDC has developed a two pronged approach of modeling datasets:

Private Sector Datasets

Established contacts with the private sector permits GEDC to obtain private sector datasets which in turn provides access to technology not readily available to the junior companies.

This partnership produces a focused approach to developing marketing tools for the Mining Sector.

Public Sector Datasets

There is a large amount of under utilized data available through the Ministry of Northern Development and Mines Geoscience Data Portal and the MNDM library. Collection and importation of this data into Datamine Studio Program will enhance the GEDC product. These additions of data will show trends between properties to entice further investment to these focus areas.

How does 3D modeling help the mining sector?

3D modeling is utilized for marketing purposes to raise the necessary funds to drive exploration and mine development.

A 3D model is a powerful visualization tool for geological formations, structure and ore body delineation. The visualization potential is only limited by the amount of data and the imagination of the people driving the model.

GEDC and the Business Community

Since 2006, GEDC has built relationships with area Junior Exploration Companies by hosting events such as 'Mining Procurement Workshop'. The goal of the Mining Procurement Workshop was to provide the small business sector and those individuals considering self-employment with relevant information to enhance the local capability to better service the mining sector.

The objectives were to:

- learn about the importance of the Mining Sector in the Greenstone Region
- network with many of the Mining Companies and Mining contractors to learn about their needs
- learn how to properly bid on product and service delivery to the Mining Sector
- learn how to promote your business products and services to the Mining Sector

A part II was held during Small Business Week in October 2008 revealing the findings revealed during the Mining Procurement Workshop. Relationships were strengthened further as GEDC continued to assist with the Greenstone business community with understanding and fulfilling the needs of the minerals sector. GEDC continued to be reminded of the importance to continue working on 3D modeling to assist in attracting other Junior Exploration Companies to the Greenstone region.

Photograph: Northern Empire Mine, 1939 (Charles Dobie)

Greenstone Mineral Development Initiative (GMDI) – funded by FedNor

GEDC successfully received additional funding from FedNor to expedite the work underway in the 3D modeling project. Eva/Summers Townships were selected as the project test site with the expectation of integrating both public and private data sets to develop a 3D geological model. At the outset, the project data compilation began and drill log data was input into Datamine Studio for the Eva / Summers Townships.

The Greenstone Belt as shown below is a '2,800-2,600 million year old greenstone belt that spans across the Ontario-Quebec border in Canada. It is comprised mostly of volcanic rocks, including ultramafic rocks, mafic intrusions, granitoid rocks, and early and middle Precambrian sediments.' (Source: http://en.wikipedia.org/wiki/Abitibi_greenstone_belt)

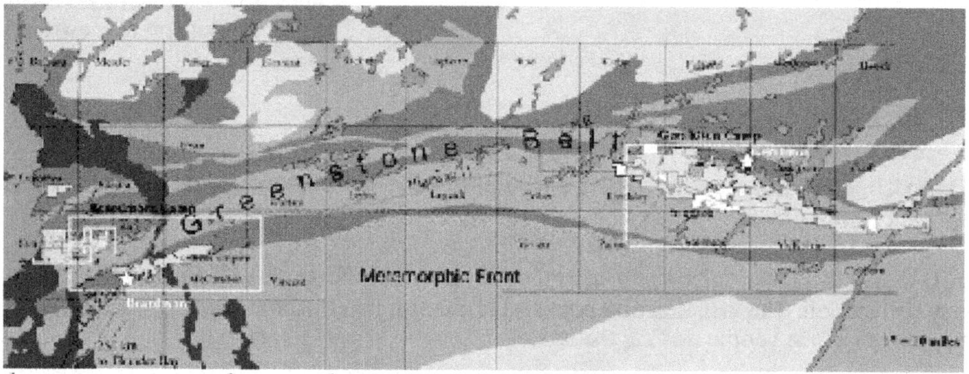

As the project progressed, new techniques were developed hand in hand with software updates which made the models more user friendly. Information can be extracted from the model to be utilized by any interested parties. Geophysics, GIS and lithology layers were also incorporated to enhance the visualization. The Northern Empire Mine was then updated to showcase the useable data in any application.

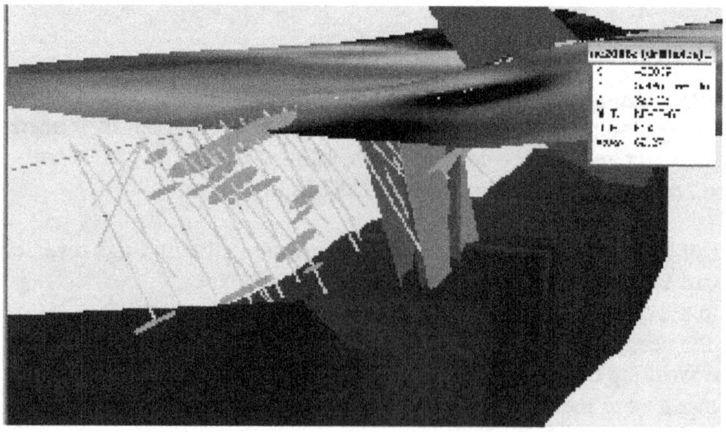

Northern Empire Mine Model (in progress)

Since the visualization is only limited by the amount of data and the imagination of the people driving the model, the GMDI project has permitted GEDC to continue to improve the 3D models and build the richness of the useable and transferable data to potentially interested companies through the newly designed marketing arm platform. GEDC believes that the OGS possesses the best location for storage of raw data files and models. As the GMDI project continues, the staff continues to be guided by a technical team comprised of public and private sector contacts. In the final phase of the GMDI, GEDC will be expected to provide FedNor with a final report outlining the recommendations of the technical team and others including the First Nations communities. The recommendations will guide the continuance of the project. It is felt by all involved that the possibilities for this application are limitless and once proven, could result in job creation for many in the Greenstone region and across the Northwest. The image above depicts the utilization of Google Earth™ as the display medium chosen by the GMDI technical team. Google Earth™ allows for direct display of the 3D data, as well as additional data which can be georeferenced. Additionally, through the interactive nature of Google Earth™ additional information is hyperlinked and raw data is made available for download. As such, the promotion and integration of 3D models can be done and the data files can be imported by anyone from anywhere in the world. As stated by Tim Twomey, Exploration Manager for Premier Gold Mines 'It is in our best interest to share all of our data because 3D data is how we make business decisions in the industry and that is how we create wealth. When we are ready to go the next phase of exploration we are going to have to raise money and we will be using the 3D tool to assist us.' The Greenstone region will thrive as the partnerships with the Mineral Sector continue to develop.

Image created using Google Earth™ Application

For further information please contact our office:

Greenstone Economic Development Corporation
Tel: (807) 854-2273
Fax: (807) 854-2474
info@gedc.ca

Developing the Future

Chapter 7 : Year 2008

Activities of Northwestern Ontario Prospectors Association

Muskeg Tours 2009

Using rocks creatively

GREENSTONE REGION STARS

NWO Mines & Minerals Symposium

A curious client, left, gets a review of the booth sponsored by Roxmark Mines Limited and Greenstone Economic Development Corporation. Looking on are Glenn Hart, Roxmark geologist; Jamie Armstrong, GEDC project analyst; and Dina Quenneville, GEDC Executive Director. Roxmark and GEDC had collaborated on a 3-D digital display of the underground workings of the old Northern Empire mine. GEDC arranged for Thunder Bay Television to interview several Greenstone-related organizations, namely, Roxmark, Kodiak, Sage, Premier Gold, Linear Metals, Freewest Resources Canada Inc., Chibougamau Drilling, and Northstar Drilling. The clips will air on Sunday, April 20, on TBT's weekly Northwest Newsweek.

Greenstone was on the map – on many maps, actually – at the Northwestern Ontario Mines & Minerals Symposium on April 8 and 9. The event took place in the Valhalla Inn, Thunder Bay.

The Northwestern Ontario Prospectors Association organized and hosted the annual event. John Halet, NWOPA President, remarked to the writer, "The year before last was very good, last year was excellent, and this year is phenomenal."

Peter Kathmann, owner of IMEXKO Contracting, tugs on a stainless steel pipe to demonstrate its firmness. A division of the company, Korex Core Racks, manufactures custom-built racks for storing core boxes. Kathmann has filled orders for many locations in Canada and the United States.

A total of 63 exhibitors manned the booths that crowded the spacious hall and spilled out into the corridor. Booths featured a range of displays from mining and industry-related businesses to government and tourism, and included one booth in the arts and crafts sector, Spirit Stone Creations from Beardmore.

The Spirit Stone Creations display stopped many spectators in their tracks. Herb Goodman and Theresa Nelson said they did brisk business in ornamental artwork and took many orders for custom cutting and polishing.

Two other Greenstone area exhibitors, aside from mining companies, were IMEXKO Contracting and Greenstone Economic Development Corporation. The picture captions elaborate on their participation.

Several Greenstone residents, current and former, attended the event. One of them, Gordon Earle, remarked, "Did you notice that the most heavily attended sessions had to do with Greenstone?" The sessions ran in an adjacent hall set up to accommodate most of the more than 400 participants in the symposium. Three companies active in Greenstone region made presentations: Kodiak Exploration Limited, Sage Gold Inc., and Premier Gold Mines Limited. Linear Metals Corporation, which participated in Greenstone's recent Mining Procurement Workshop, also presented.

Each of the six sessions had a succession of presentations, each presentation lasting about 20 minutes. IMEXKO Contracting sponsored Session D, which included a presentation titled "Exploration Opportunities in the Vicinity of the [James Bay] Lowlands". The company has supplied junior miners with core racks in the McFauld's Lake Area.

One of the highlights of the event occurred at the NWOPA Awards Luncheon on April 9. The Bernie Schnieders "Discovery of the Year" Award went to Kodiak Exploration for their Hercules Project in the Beardmore-Geraldton gold camp.

In accepting the award, Bill Chornobay, Kodiak's President, thanked several enablers, including Greenstone's munipality, local First Nations, outfitters, suppliers, prospectors, and other stakeholders.

Not since the 1930's has Greenstone region received such attention in a popular and widely respected mining forum.

- Times Star

NORTHWESTERN ONTARIO PROSPECTORS ASSOCIATION

TO REPRESENT AND FURTHER THE INTERESTS, SERVE THE NEEDS, AND SUPPORT THE ASPIRATIONS OF THE INDIVIDUAL PROSPECTORS IN NORTHWESTERN ONTARIO

Visit NWOPA'S Website at http://www.nwopa.net/ to obtain more information on our annual NW Ontario Mines & Minerals Symposium and other NWOPA Events

Chapter 8 :
Year 2008

Activities of Northwestern Ontario Associated Chambers of Commerce

Muskeg Tours 2009

A sleeping giant awakens
in the Northwest

NOACC SPRING MEETING

Greenstone Hosts Business Leaders

"It has been one of the most successful meetings in our history in terms of information and presentations," Barry Streib told the Times Star. Streib is President of the Northwestern Ontario Associated Chambers of Commerce (NOACC).

NOACC met in Geraldton from Thursday evening to Saturday noon, April 24-26. The venue for workshops and meals was the French Club. The Friday night dinner took place in the Curling Club lounge.

Streib was describing the powerful lineup of speakers at NOACC's semi-annual general meeting. He cited the example of NOACC's new relationship with the Ontario Mining Association, in recognition of the role mining exploration is playing throughout the Northwest. The Spring Meeting had a mining theme, but topics ranged from mining to community infrastructure to government initiatives in the North.

Dave Barker calculated that about 30 registered delegates participated in the series of workshops. Barker recalled that the last Spring Meeting in Geraldton occurred in 1992, for he chaired that session.

Peter Kathmann, President of Geraldton & District Chamber of Commerce, said that he had received many comments praising the session. The local Chamber hosted the NOACC event. "On a scale of 1 to 10," he said, "I give it 11".

Mining Grows a Community

"The Northwest has a huge potential . . . I'm pretty positive on it." Chris Hodgson was addressing the Northwestern Ontario Associated Chambers of Commerce on April 25 in Greenstone.

Hodgson, President of the Ontario Mining Association (OMA), devoted most of his presentation to a report commissioned by OMA and the Ministry of Northern Development and Mines. The latest report, one of a series of semiannual reports on the contributions of mining in Ontario, studies the economic impacts of a "representative" mine in Ontario. A fictional nickel-copper mine represents a typical mine. The report can be read online at www.oma.on.ca. Click on "Economics".

Karen MacDougall holds a miner's gas torch, one of the many artifacts on display in the Curling Club lounge. The interior of the tent, the prospector's home in the field, cleverly suggests that the miner has just left for a few moments, leaving his clothes to dry by the propane stove and his book open where he left off reading.

Overall, a typical mine in its construction phase adds about $140 million to Ontario's gross domestic product (GDP), and for each year of production it adds almost $280 million. All dollar figures quoted use 2005 currency values.

The impacts on the economy are staggering. The "representative" mine results in 480 direct jobs, and another 1,103 indirect (i.e., mining-related) jobs. The induced employment adds another 697 jobs, for a grand total of 2,280 jobs. Induced employment results from the spending of the wages and salaries of the direct and indirect job holders. In terms of the serviced community within one hour's drive of the mine, employment increases by 1,519 jobs, which is 67% of the total province-wide impact.

As for the dollar impact, the operating mine will produce, each year, $150.5 million as direct GDP, $81.4 million as indirect GDP, and another $45.9 million as induced GDP, for a grand total of $277.8 million. In terms of the local community, the GDP impacts are $219.5 million, i.e., 79% of the total.

Chris Hodgson, President of OMA, and Barry Streib, President of NOACC, at the French Club.

Other economic impacts not included in the figures above are: charitable donations/investment in community development (such as medical and recreation facilities); spending of pension income; further mining exploration and development; benefits to Aboriginal communities; mine reclamation and rehabilitation; and development of downstream industries such as smelting and refining.

Another upside for the taxpayer is that the representative mine generates millions of dollars for all levels of government, totaling $83.8 million annually, of which municipal government receives $10.6 million.

In response to a question, Hodgson said that Canada is recognized internationally for its expertise in two areas. Speaking tongue in cheek, he said, "Internationally we're known for hard rock mining and women's hockey."

He concluded the presentation by quoting a leading environmentalist who stated that mining can be compatible with respect for the environment.

The Rosehart Report

Dr. Robert G. Rosehart addressed the Northwestern Ontario Associated Chambers of Commerce on April 25 during dinner at the Geraldton Curling Club. His official report to the Government of Ontario was released earlier this year.

The Rosehart Report deals with issues in Northwestern Ontario (NWO), and recommends actions to stabilize the economy, to build capacity for the "New Economy", and to invest in the region's future.

In one of his opening remarks, Rosehart pointed to job losses across Ontario. He stated, "The challenge we face that they [Southern Ontario] don't face is that in Southern Ontario the jobs get replaced."

There are two significant population trends. Between 2006 and 2021, the population of Ontario will grow 33%. In NWO, the population will decrease 7%. At the same time NWO will experience a growth in its Aboriginal population.

Three leading figures pose before the prospector's tent that illustrated the mining theme of the Spring Meeting: Peter Kathmann, President, GDCC; Dr. Robert Rosehart, dinner speaker; Barry Streib, President, NOACC.

The Report makes 47 recommendations. "There's three of them in there," said Rosehart, "and I don't say which three, that add up to a billion dollars each." He pointed out that NWO is strategically located between Western and Eastern Canada. "How do we get people thinking of Northwestern Ontario as a strategic piece of real estate?" he asked. NWO must somehow get the attention of Queen's Park.

He said that the recent Ontario budget incorporated three of the Report's recommendations: geophysical mapping to assist mining exploration ($20 million over 4 years); acceleration of the business education tax rate cut; and a centre in Thunder Bay for research and innovation in the bio-economy.

Rosehart described the Report as a "road map", a plan, "something you can thump the table with to create action".

One recommendation refers to developing all-weather roads to the Far North. It must be, he said, part of a vision of nation-building, something like the now defunct Roads to Resources plan of decades past. Referring to the plight of fly-in communities, he said, "I don't think you can drive any economy when you're paying $7.00 a litre for milk." He envisioned the recommendation's implementation over 30 years.

"The objective in the long-term is to have a prosperous region," he said. He suggested that even the region's aesthetics need improvement. Visitors and potential investors can drive from the Manitoba border to White River and see from 150 to 200 boarded-up buildings. At one point he commented on the public washroom facilities along the highways which are locked up for the winter, posing serious inconvenience to travelers, even local ones.

He remarked that the forest industry does not feel it has "a champion" in government. The Report recommends a biomass power generation facility, and a reduction in stumpage fees when transportation costs from stump to mill are excessive.

He touched briefly on tourism. "If we really want to encourage American tourists, we have to learn to be friendlier to them." He alluded to Canada's attitude to D.U.I. convictions, which is much more severe than the American's.

Referring to the jailing of KI First Nation residents in its dispute with a mining company, Rosehart emphasized the need for dialogue. Even when the Mining Act is updated, and the "duty to consult" section is supposedly improved, the section will still need to be tested and defined in the court system, and that will take years.

With regard to Aboriginal schools , he said, "First Nation students get about 33 cents on the dollar," compared to non-Aboriginal students. The funding provided by Indian and Northern Affairs Canada should be supplemented with Provincial funds, along with other initiatives.

In winding up his presentation, Rosehart stated, "Particularly with respect to the big ticket items, they'll never happen unless you have the two senior levels of government working together, the Canadian and the Ontario governments."

He called for leaders to step forward. "At some point leadership is about making a decision and getting on with it . . . If the Northwest is going to get special consideration, the North is going to have to fight for it."

Rosehart sat down to resounding applause from the delegates and guests.

Speaking later to the Times Star, Rosehart said that he was quite willing to travel again to Greenstone to speak to other groups, including the general public.

GCFC Launches New Site

On April 26 Tara Gallant outlined a new Web-based service to the semi-annual general meeting of the Northwestern Ontario Associated Chambers of Commerce. Gallant is GIS Supervisor of GEOMA, a division of Geraldton Community Forest Corporation.

GCFC, a nonprofit community corporation since 1994, has diversified its services to include geographic information services (GIS). The day before, the Minister of Northern Development & Mines, Michael Gravelle, had announced funding for another of GCFC's programs, Community GIS, which aids investment readiness in the communities of Greenstone, Red Rock, Nipigon, Schreiber, and Ear Falls. At the same time, he helped to officially launch the first GEOMA project.

"One of the most exciting projects we have on the go today is NOmap," said Gallant in her slide presentation. The name NOmap is suggestive of Northern Ontario mapping service. Businesses, organizations, and communities across the North are invited to register at NOmap.ca.

The annual fee is $150, but the first six months come free. NOmap is a tool to locate places, geographical features, and short text information.

Using the interactive feature, an Internet surfer – for example, a potential tourist or a business supplier – can activate different layers of information, which are then displayed on sectional maps of Northern Ontario. One can zoom in or out. Layers include such information as lakes, fish species in those lakes, roads, communities, organizations, businesses, accommodations, local attractions, snowmobile and hiking trails, and canoe routes.

Partners in the GEOMA project are the Northern Ontario Heritage Fund Corporation and ESRI, a international firm with an office in Sudbury. The firm supplies the GIS software.

- Times Star

Tina Callant
GIS Supervisor
of GEOMA

PROSPECTS NORTH CONFERENCE
NOACC Engages Mining Industry

A mining conference took place at the Travelodge Hotel Airlane on Thursday and Friday, October 23 and 24, an initiative of the Northwestern Ontario Associated Chambers of Commerce and the Ministry of Northern Development and Mines. As far as anyone could recall, it was the first regional conference in the Northwest to engage mainly senior (major) mining companies, as opposed to junior exploration companies.

Chris Hodgson gave the keynote address at the dinner on Thursday evening. Hodgson, President of the Ontario Mining Association, addressed approximately 160 people under the rubric "A Partner in Prosperity Building". He stated that the OMA's 58 members including operating mines, contractors, suppliers, and others associated with the industry.

Barry Streib (left), President of NOACC, poses with Chris Hodgson, President of OMA, in front of the conference poster.

The value of mineral production in Ontario in 2007 reached $10.7 billion, producing 31% of Canada's metallic minerals and 22% of the nonmetallic. The year 2007 also saw a record $502 million in exploration and related expenditures.

The Toronto Stock Exchange, he said, is the mine financing capital of the world, having raised 35% of the total world equity capital in 2007. It is also home to 57% of the world's listed mining companies.

Hodgson pointed out that the average operating mine purchases 43% of its goods and services "within a mile of the headframe", or at worst within an hour's drive of the mine.

The reality today, he said, includes the uncertain state of the current commodity markets, and fact that so many mining companies operate on a global scale, and invest only where they are welcome. However, he continued, "Strong demand from growing nations will keep metal prices healthy for years to come."

Hodgson described mining as "an energy-intensive industry" Energy prices continue to rise even though the industry has made great strides in achieving energy efficiency. Co-generation is one solution. Therefore, he said, "A megawatt saved should be as valuable as a megawatt generated."

Hodgson alluded to the Ontario government's initiative to modernize the Mining Act, stating that it was imperative to strike a balance between social aspirations and a healthy industry. As for Premier McGinty's initiative to develop a land use strategy for the Far North, he said, "If we get this right, it would be a marvelous achievement."

Hodgson referred to the problem of growing the industry's workforce. "If employers have difficulty now in recruiting, we'll have a disaster in five years." He suggested some directions in which to look: youth, immigrants, and women. "The group that will solve our problem," he said, "is women." Women represent 13% of the industry's workforce, but the industry needs at least 25% representation. "People think of workers at the face of the rock [in a mine]. It's an image problem." He suggested that there are many other opportunities in mining.

In conclusion, Hodgson emphasized that mining has a bright future, not the least because industry, governments, NGO's (specifically environmental groups), and First Nations appear to be pulling in the same direction.

The writer conducted interviews with some key people. A Co-Chair of the Conference stated, "We wanted to create a learning opportunity for NOACC members as well as people in the mining and business community." Ian McCormack added, "We felt that First Nation communities and individuals should have and can have opportunities in that industry." NOACC's goal was to present "strategic perspectives" rather than to debate issues.

Co-Chair Barry Streib envisaged a series of annual conferences. Streib, President of NOACC, said, "In our next phase, I'd like to see a more global perspective." This phase would involve invitations to the mining sector all across Canada as well as beyond our country's borders. He stated that this conference, having 150 registered delegates, had met its attendance target.

One of the presenters, Terry Waboose, Deputy Grand Chief of Nishnawbe Aski Nation (NAN), spoke to the writer by phone. "Part of our mandate and job," he said, "is to educate . . . the broader community, including industry and governments and municipalities." His goal as presenter was to give the perspective of First Nations on their relationships to government and the mining sector. He remarked that he had received positive feedback: "My message was well received."

Outside the conference proper, the Province of Ontario and Matawa First Nations held a press conference late Friday morning in order to sign a letter of intent. Michael Gravelle, Minister of Northern Development and Mines, represented the government. The letter documented the common goal of developing a protocol on traditional Matawa lands. Speaking to the writer, Waboose (NAN) commented, "It is a milestone because it opens up that dialogue with government for those First Nations that did sign."

Speaking to the writer, Streib (NOACC) had previously remarked, "We're all partners in this thing whether we think we are or not."

The Prospects North conference, an intiative of NOACC and MNDM, was also supported by the Thunder Bay Chamber of Commerce and OMA.

Canada's Leading Role

"When the world thinks of mining, they think of Canada first," said Ian McCormack. The General Manager, Thunder Bay Operations, of Wardrop Engineering Inc., was addressing Northwestern Ontario's first annual mining conference on October 24.

A fellow presenter, Dr. Christine Kaszycki, espoused much the same view. Kaszycki, Assistant Deputy Minister with the Mines and Minerals Division of MNDM, pointed out Ontario's ranking in the world's non-fuel mineral production: 3rd in platinum, 4th in nickel, 7th in cobalt, and 11th in gold.

McCormack described Wardrop as one of the top engineering consulting companies in the world, with 1,100 employees in 17 offices worldwide. "We compete all the time," he said, "with the big brand names in engineering consulting, going head to head." They work with senior mining companies as well as junior exploration companies, "the risk-takers", as he described them.

Wardrop deliberately located their head office on Bay Street, "right in the heard of the investment community in mining and minerals". They provide services to regional companies such as Goldcorp in Red Lake and the Williams Mine at Hemlo. "Ontario is the largest producer of metals in Canada," he asserted, "and we have the largest mineral deposits on the face of the earth."

Referring to the current economic slowdown, McCormack alluded to the annual Policy Potential Index published by the Fraser Institute. The Index is the result of survey of mining companies around the world, ranking countries and regions in 14 different policy areas according to their investment attractiveness.

A glance at the Index reveals that in 2007-08, Quebec was the most highly rated single jurisdiction in the world, with Alberta rated 4th, and Manitoba, 5th. Ontario, however, rated only 18th. Still, the average score of Canada and the provinces placed them ahead of Australia (another big mining country) and the United States.

McCormack said, "The point is, our clients read this stuff." He added later, "We need to strive to maintain our leadership position worldwide."

He suggested that the economic situation for the next 12 months would likely be "very painful", but it would improve, although it may take from 6 to 8 months to feel the difference. "Our clients don't base their project decisions on the daily ups and downs of the market," he said. Wardrop recommends to their clients that they make long-term commitments and "lock up their resources" as they weather the slump.

In her presentation, Christine Kaszycki of MNDM highlighted the good news in Ontario's mining sector in the last while. Using figures from 2007, Kaszycki cited mineral production of $10.7 billion, $1.1 of that in Northwestern Ontario (NWO). The direct employment figure was 15,000, with 2,500 of that in NWO. Expenditures in exploration were $502 million, with $190 in NWO. There were over 1,000 companies operating in the mining services sector, 170 in NWO.

Currently, there are some 800 exploration projects in Ontario, 350 of those in NWO. The expenditures on exploration by junior miners now equal those of the seniors (majors). In 2007, of 308,000 active claims in Ontario, Thunder Bay led all other Mining Divisions with a total of 84,000.

Kaszycki echoed McCormack's assessment of the current economic situation. She referred to the drop in metal prices globally, and to the rise of the industry's costs. She stated that investment money is not flowing as before to the higher risk sector, i.e., the juniors. She asserted her faith in the long-term soundness of the industry. As Chris Hodgson and Ian McCormack did in their presentations, she pointed to the shortage of skilled labour.

Still, Kaszycki saw reasons for optimism. There is Ontario's mineral potential and it's well developed infrastructure. In policy areas, there is the modernization of the Mining Act (which addresses land tenure, a major factor in attracting investment), the efforts to enhance the participation of First Nations, and her Ministry's investments in such areas as geological mapping.

In his opening remarks, McCormack had stated, "Being Canadian in the mining industry around the world means being the best." It also means, all presenters agreed, staying the best.

Left: Ian McCormack of Wardrop Engineering

Right: John Mason, North-west Regional Manager, & Dr. Christine Kaszycki, DM of MNDM

First Nations Ready to Participate

At the region's first annual mining conference in Thunder Bay on October 24, one geologist commented to the writer that he was surprised and delighted by the representation of First Nations both as presenters and as delegates.

Scott Jacob, Chief of Webequie First Nation, opened his presentation with the remark, "We want a share of the experience of intensive exploration activity within our own community." He envisaged partnerships with the private sector. "The parties must share," he said, "the same vision, a vision of economic prosperity, that is beneficial to both parties."

In his own presentation, the Deputy Grand Chief of Nishnawbe Aski Nation (NAN) emphasized the responsibilities of both government and the mining sector. Terry Waboose of NAN represented communities in James Bay Treaty 9, and the Ontario sector of Treaty 5, which also embraces Manitoba.

Chief Jacob's community is located in the vicinity of the so-called Ring of Fire in the James Bay Lowland, also known as the McFauld's Lake area, where approximately 40 juniors are exploring for base metals and diamonds. "That comprises about 2.5 million hectares of prime traditional lands between First Nations in the area," said Jacob.

Webequie FN has a 20% interest in Cyr Drilling International in the form of Cyr Drilling Ontario Inc., and nurses a goal of finding work across the whole of Ontario. The community is a base for helicopters and fixed wing aircraft, and is a supply base for equipment and supplies in the Ring of Fire. It also provides field workers for exploration companies. It is necessary, said Jacob, to increase opportunities for employment and for training, and to participate in the benefits of the exploration boom. He added, "In spite of many meetings, time, and expenses, negotiations have not been satisfactory."

One of Webequie FN's goals is to work in concert with other FN's, for there is no intention to compete with them for jurisdiction on the traditional lands that have overlapping claims.

Deputy Grand Chief Waboose said that NAN represents 49 communities across 55 million hectares, two-thirds of Ontario's land base. He acknowledged the problems that the mining industry faces, but insisted that the industry needs to recognize the FN perspective.

Deputy Grand Chief Terry Waboose, NAN, believes his message was well received.

Being signatories to treaties, FN's have a direct government-to-government relationship with the Crown. The context of this relationship, he said, is outlined in Canada's Constitution Act of 1982, Section 35 (1). He said that the Province of Ontario tends to downplay this relationship, insisting on looking first at the economic impacts of mining on the whole of Ontario, and only secondarily at the impacts on FN's.

"The system needs to get right," Waboose said, "before we can attract real investment opportunities." NAN believes in the "sustainability approach" and in benefits to FN's.

As things stand, he said, a company has the responsibility to notify FN's of its proposals, but it is still the duty of the Crown to consult with them. This is particularly necessary where a company's program may breach or even annihilate an aboriginal right.

FN's, he said, are usually quite willing to enter into exploration agreements with the private sector which detail such things as benefits, mitigation of impacts, and levels of participation. However, these business agreements do not excuse the Crown from fulfilling its duty to consult. The Crown must protect and accommodate aboriginal rights.

The concerns of FN's, therefore, extend far beyond the Mining Act now being revised. The concerns include benefits such as revenue-sharing (e.g., taxes, royalties).

After the presentations by Jacob and Waboose at the Travelair Hotel Airlane in Thunder Bay, the Ministry of Northern Development and Mines held a special press conference outside the conference venue. Michael Gravelle, Minister, signed a letter of intent with Matawa First Nations (some of whom participate in NAN) to establish a protocol for government-to-government consultation. Matawa provides technical services and program delivery to 10 FN's in NAN, including Webequie.

In his closing remarks to the conference, Chief Jacob emphasized that FN's want to and need to participate in the current exploration boom: "We all suffer from poverty, unemployment, and other social ills . . . and our members expect us to do what we can to help."

Mining Companies on Board

At the regional conference in Thunder Bay on October 24, mining companies outlined their operations. Two presentations are discussed below.

Abraham Drost, President and CEO of Skybridge Development Corp., stressed the important role of junior exploration companies. He said that Canada (via the Toronto and Vancouver stock exchanges) is the world's centre for venture capital upon which juniors depend. Juniors do the systematic evaluations of mineral properties listed on these exchanges.

Abraham Drost stands by the Skybridge booth at the conference.

Skybridge has been involved in many mineral discoveries and development projects in North America, with considerable success in the Arctic. Its Blue Caribou Project in Nunavut is a promising copper-molydbdenum occurrence. Ewan Downie, Chairman and Director, is also President and CEO of Premier Gold Mines Ltd., now exploring in Geraldton.

Andrew Baumen of Hemlo Mines spoke on the Williams mine and the David Bell mine near Marathon. As Technical Service Manager, he demonstrated a grasp of the economics of running a gold mine.

Drost (Skybridge) explained that junior companies are traditionally part of the supply chain for senior companies. In fact, seniors typically grow by acquiring juniors. The highlight of his presentation was a listing of juniors with offices in Thunder Bay, a grand total of 79.

Drost enumerated current issues in the industry, duplicating most of the ones listed by earlier presenters, such as the shortage of skilled labour, aboriginal land claims, government policies, and the poor image of the industry as a whole.

In the current economic slowdown, he saw silver linings. For example, he pointed out the difference in the price of "paper gold" (the market price usually quoted by the media) and "physical gold" (gold bullion). Bullion is priced much higher than paper gold. His recommendation?: "In short, buy gold."

As for the slump in commodity prices, "This sell-off [of mining company shares] in the market has put cash on the sidelines," he said. The money did not vanish; what is happening is "an evaporation of apparent value", and the business cycle will see an inevitable rise in share prices.

Andrew Baumen described Hemlo Mines as a joint venture of Barrick Gold Corp. and Teck Cominco. They employ 600 people in two underground mines and one open pit operation, producing 260,000 ounces per year. These are low-grade operations (averaging 3 grams per tonne), with the resources to operate another 3 to 5 years.

Operation costs include the cost of hydro, diesel fuel, blasting supplies, and above all, labour. The workers, including contractors, consume about 50% of the $200 million annual budget. About 20% is spent nationwide, 40% in Ontario, and 40% ($80 million) locally.

Baumen's presentation suggested to the audience the kind of prosperity that may visit other communities if any of the 79 juniors operating out of Thunder Bay achieve a resounding success.

- *Times Star*

From L to R, the open pit operation, the Williams mine, the Newmont mine (closed), and, bottom right, the David Bell mine. MNDM Photo.

On October 24, just outside the conference venue, the representatives of Matawa First Nations sign a letter of intent with Michael Gravelle (right), Minister for MNDM. Third from right is Chief Scott Jacob, a conference presenter. The letter documented the common goal of developing a protocol for consultation.

THUNDER BAY HOSTS THE FIRST ANNUAL PROSPECTS NORTH MINING CONFERENCE

A mining conference took place at the Travelodge Hotel Airlane on Thursday and Friday, October 23 and 24, an initiative of the Northwestern Ontario Associated Chambers of Commerce (NOACC) and the Ministry of Northern Development and Mines (MNDM).

Ian McCormack, Co-Chair of the Conference, stated to the writer, "We wanted to create a learning opportunity for NOACC members as well as people in the mining and business community." He added, "We felt that First Nation communities and individuals should have and can have opportunities in that industry." NOACC's goal was to present "strategic perspectives" rather than to debate issues.

Chris Hodgson gave the keynote address at the dinner on Thursday evening. Hodgson, President of the Ontario Mining Association, pointed out that the Toronto Stock Exchange is the mine financing capital of the world, having raised 35% of the total world equity capital in 2007. It is also home to 57% of the world's listed mining companies.

Hodgson set the tone of the conference, enumerating the challenges facing the industry, and suggesting reasons for optimism. The reality today, he said, includes the uncertain state of the current commodity markets, and fact that so many mining companies operate on a global scale, and invest only where they are welcome. However, "Strong demand from growing nations will keep metal prices healthy for years to come."

One of the challenges is growing the industry's work force. "The group that will solve our problem," Hodgson said, "is women." Women represent 13% of the industry's workforce, but the industry needs at least 25% representation. "People think of workers at the face of the rock [in a mine]. It's an image problem." He suggested that there are many other opportunities in mining.

On Friday, two representatives of First Nations (FN's) addressed the conference. Scott Jacob, Chief of Webequie FN, opened with the remark, "We want a share of the experience of intensive exploration activity within our own community." He envisaged partnerships with the private sector. "The parties must share," he said, "the same vision, a vision of economic prosperity, that is beneficial to both parties."

Chief Jacob's community is located in the vicinity of the so-called Ring of Fire in the James Bay Lowland, also known as the McFauld's Lake area, where approximately 40 juniors are exploring for base metals and diamonds. "That comprises about 2.5 million hectares of prime traditional lands between First Nations in the area," said Jacob.

Jacob outlined the many ways in which his community is participating in the mining exploration, but added, "In spite of many meetings, time, and expenses, negotiations have not been satisfactory."

In his own presentation, the Deputy Grand Chief of Nishnawbe Aski Nation (NAN) emphasized the responsibilities of both government and the mining sector. Terry Waboose of NAN represented communities in James Bay Treaty 9, and the Ontario sector of Treaty 5, which also embraces Manitoba.

Being signatories to treaties, FN's have a direct government-to-government relationship with the Crown. "The system needs to get right," Waboose said, "before we can attract real investment opportunities." NAN believes in the "sustainability approach" and in benefits to FN's.

FN's, he said, are usually quite willing to enter into exploration agreements with the private sector which detail such things as benefits, mitigation of impacts, and levels of participation. However, these business agreements do not excuse the Crown from fulfilling its duty to consult. The Crown must protect and accommodate aboriginal rights.

Ian McCormack spoke about the role of his company, Wardrop Engineering Inc. McCormack, General Manager, Thunder Bay Operations, stated, "When the world thinks of mining, they think of Canada first." However, he foresaw the next 12 months as likely "very painful", but it would improve, although it may take from 6 to 8 months to feel the difference. "Our clients don't base their project decisions on the daily up's and down's of the market," he said. Wardrop recommends to their clients that they make long-term commitments and "lock up their resources" as they weather the slump.

A fellow presenter, Dr. Christine Kaszycki, stressed the good news in the mining sector in recent years. Kaszycki, Assistant Deputy Minister with the Mines and Minerals Division of MNDM, used figures from 2007. She cited mineral production of $10.7 billion, $1.1 of that in Northwestern Ontario (NWO). The direct employment figure was 15,000, with 2,500 of that in NWO. Expenditures in exploration were $502 million, with $190 in NWO. Currently, there are some 800 exploration projects in Ontario, 350 of those in NWO.

Kaszycki stated that investment money is not flowing as before to the higher risk sector, i.e., the juniors. She asserted her faith in the long-term soundness of the industry, and referred to recent steps her Ministry is taking, such as modernization of the Mining Act.

A number of mining industry representatives discussed their companies' roles. Among them, Abraham Drost, President and CEO of Skybridge Development Corp., stressed the important role of junior exploration companies. The highlight of his presentation was a listing of juniors with offices in Thunder Bay, a total of 79.

Andrew Baumen, Technical Service Manager, described Hemlo Mines (near Marathon) as a joint venture of Barrick Gold Corp. and Teck Cominco. They employ 600 people in two underground mines and one open pit operation, producing 260,000 ounces of gold per year. They spend $200 million on goods & services per year, 40% of that locally.

Baumen's presentation suggested to the audience the kind of prosperity that may visit other communities if any of the 79 juniors operating out of Thunder Bay achieve a resounding success.
Speaking to the writer, Co-Chair Barry Streib envisaged a series of annual conferences. Streib, President of NOACC, said, "In our next phase, I'd like to see a more global perspective." This phase would involve invitations to the mining sector nationally as well as internationally. He stated that this conference, having 150 registered delegates, had met its attendance target.

- Northern Ontario Business

The co-chairs pose with the Deputy Grand Chief of NAN. L to R, Ian McCormack, Terry Waboose, and Barry Streib.

NOACC

Northwestern Ontario Associated Chambers of Commerce

Vision

To be the voice and active link in the formation of public policy contributing to the economic development and quality of life in Northwestern Ontario.

Visit

www.noacc.ca

Northwestern Ontario Associated Chambers of Commerce

Mission

The Northwestern Ontario Associated Chambers of Commerce (NOACC) is the active voice of business in Northwestern Ontario. Through the cooperative partnerships and collaborative efforts of our local Chambers and businesses, we lobby the government to promote active and relevant solutions to Northern issues.

We are committed to the preservation and development of our unique communities to ensure a healthy economic future and a superior quality of life for our member Chambers, regional business districts and the citizens of Northwestern Ontario.

Vision

To be the voice and active link in the formation of public policy contributing to the economic development and quality of life in Northwestern Ontario.

- Northwestern Ontario Associated Chambers of Commerce (NOACC) represents every active Chamber of Commerce from Marathon to the Manitoba border.
- Our membership includes over 2000 businesses and 7 of the largest corporations in the Northwestern Ontario region.
- We are dedicated to continued economic growth and development in our communities and are among the most influential advocates for business in the province.
- NOACC works in co-operation on common issues with regional, provincial and national organizations to advance the goals of our Membership through the formulation of public policy.

We are actively involved in the identification and resolutions of issues and concerns affecting the sustainable development of Northwestern Ontario.

Chapter 9 :
Year 2008

Activities of Roxmark Mines Limited

Muskeg Tours 2009

Shaft at the old Northern Empire

Helping to assure a prosperous future for the Beardmore-Geraldton Greenstone Belt.

Working alone and in partnership with Premier Gold, Roxmark is developing 12 gold properties and one molybdenum resource.

Our plan is to re-open the Northern Empire Gold Mine and to advance others which produced nearly two million ounces of gold, taking advantage of a modern mill and other existing infrastructure to speed development and reduce costs.

Priorities for 2009 include exploration of promising targets at the Northern Empire and on the neighbouring Nortoba Tyson property.

Over the coming years, we hope to contribute to economic development and greater employment in the Beardmore-Geraldton Greenstone region.

Roxmark's 200 TPD Northern Empire Mill

ROXMARK MINES LIMITED

Mine Office
P.O. Box 250 • Beardmore ON
Head Office

ROXMARK MINES' 2009 GOLD EXPLORATION AND DEVELOPMENT PROGRAM TO FOCUS ON PROPERTIES CLOSEST TO BEARDMORE MILL

2009 is going to be a busy time for Roxmark Mines as this pioneering Company, which helped spur recent gold activity in the Beardmore-Geraldton area, proceeds toward reopening the historic Northern Empire Gold Mine.

Roxmark's 2009 field program, including a minimum of 5,000 metres of diamond drilling funded by $1.35 million in new flow-through financing, will concentrate on proving additional gold resources at the Northern Empire and on the neighbouring Nortoba-Tyson property, both of which are on the doorstep of the Company's Northern Empire mill in Beardmore.

The Beardmore Camp program, which will also include geological mapping, channel sampling and other surface work, is based on interpretation of a recent High-Resolution Aeromagnetic (HRAM) Survey completed by Firefly Airborne Surveys on Roxmark lands. A total of 1,983 line-km of airborne magnetic survey with a spacing of 50 metres was covered by fixed wing aircraft and G-822A Cesium Magnetometer.

The survey outlined several anomalies that warrant follow-up using knowledge gained from previous drilling and channel sampling.

In November 2008, Roxmark issued an NI 43-101 compliant technical report on the Northern Empire by independent geologist Peter Bevan which outlined 86,652 tonnes grading 10.70g.Au/t containing 29,807 ounces of Indicated Gold Resources; and 64,748 tonnes grading 9.95g.Au/t containing 20,719 ounces of Inferred Gold Resources, at shallow depths over a limited drilled strike length of the Contact and Power Zones. Previous news releases described both resources and potential in detail.

Only the top 100 metres of the Contact Zone have been drill tested, in a system known to host good gold mineralization to a depth of more than 600 metres. According to the report, historical production on veins that were open at depths of 2,075 feet provides conceptual evidence that substantial additional potential exists in the Contact Zone and parallel structures far below the diabase sill.

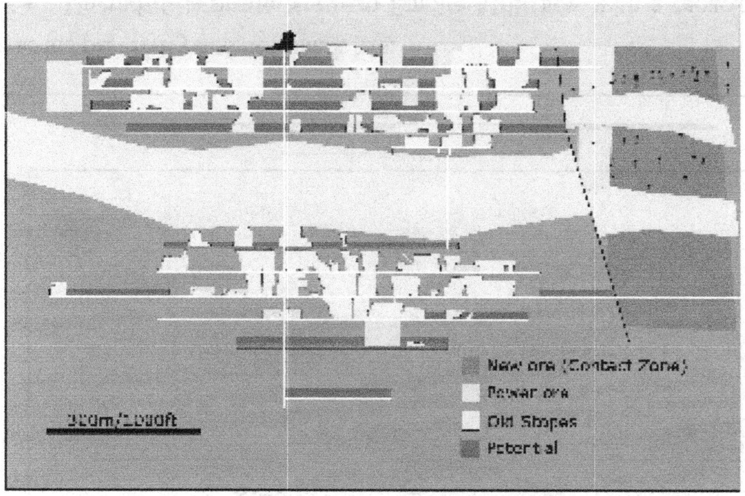

Northern Empire Historical Reserves	Tons	Grade (Oz/ton)	Oz Au
Contact Zone East (above 2nd level)	45,627	0.43	19,425
Power Zone East (above 2nd level)	20,342	0.56	11,355
Total Probable Reserves *not NI 43-101 compliant	65,969	0.47	30,780

"We remain convinced that early underground development at the Northern Empire will be economically attractive," said Monir Younan, Roxmark's President. "Our objective for 2009 is to explore other zones on the property which could add to our resource base. Previously announced 2008 exploration results on our Nortoba-Tyson gold/molybdenum property were also encouraging," Younan added. "However, drilling was limited to a few holes on which we will be following up, based on our interpretation of last year's data."

"Overall, we are optimistic about prospects for our properties in the Beardmore and Geraldton Camps," Younan added. "In addition to natural resources, we believe the human resources in the area represent an important asset that will assure a bright future."

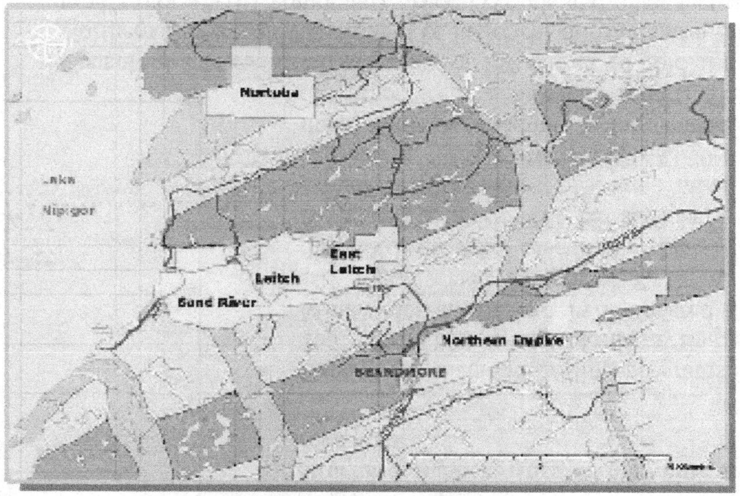

Premier-Roxmark Joint Venture

Roxmark is also an active participant (with a 30% carried interest) in The Hardrock Project, a joint venture with Premier Gold Limited to develop a number of Geraldton Camp gold properties. In 2009, as operator, Premier will be spending an estimated

$8 million to carry out a three-drill, 50,000-metre program to delineate resources on Project lands directed at establishing thresholds for early commercial development by both open-pit and underground mining. Expectations are high.

About Roxmark

Roxmark Mines Limited has been a leader in gold and molybdenum exploration and development in the historically prolific Geraldton-Beardmore gold mining area. In recent years, Roxmark has generated cash flow from bulk-sampled gold and molybdenum processed at its fully-permitted mill and has the advantage of infrastructure from six formerly highly productive gold mines located on its properties. These mines previously produced nearly two million ounces of gold from high grade ore but were closed primarily due to dramatically lower gold prices at the time and to boundary issues, since eliminated by Roxmark.

Five Years of Accomplishment for Roxmark

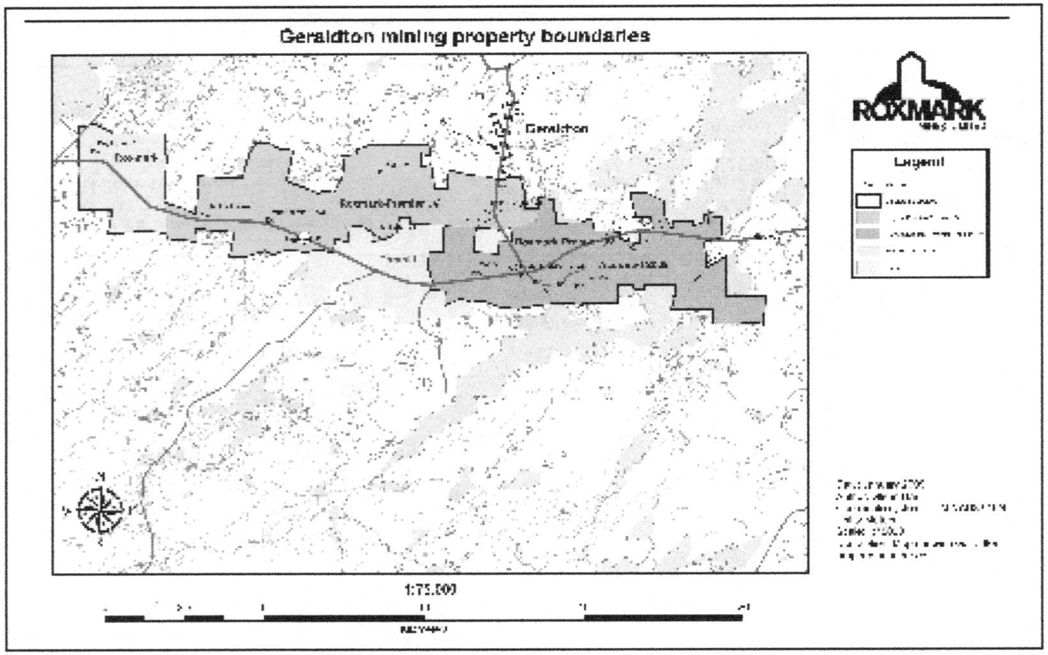

Over the past five years, Roxmark has:

- Upgraded its onsite Northern Empire Mill (now with a capacity of 200 tons per day, but expandable to 500 tons per day) to enable processing of both gold and molybdenum, adding a gravity concentrator and new flotation circuit and dryer to maximize efficiency.
- Surface bulk-sampled 800 tons of gold from the East Leitch property and more than 2,000 tons of molybdenum ore from the Nortoba-Tyson property, and processed both through its mill. Roxmark has also worked to establish the commercial viability of molybdenum production at Nortoba-Tyson, taking another bulk sample at surface for future processing in 2008, using production techniques designed to minimize dilution of ore delivered to the mill.
- Conducted an 1,865-metre diamond drilling program on the Nortoba-Tyson property which further delineated and extended the strike length (717 metres) and depth (1,000 ft.) of the molybdenum mineral deposit, and assessed the structural characteristics of the veins and rock wall.
- Carried out diamond drilling programs in 2007 and 2008 at the Northern Empire Mine which established, in an updated 43-101 report dated November 8, 2008, a threshold of gold resources which the Company believes will be sufficient to support re-opening of the mine—beginning with the 3,500-4,000-ton underground bulk sample recommended in the report. In a previous report, SRK Consulting had estimated resources of 150,000 oz of gold at the Northern Empire on the strength of historical drilling and a 3-D model (not 43-101 compliant). Based on this and other factors, Roxmark management believes it is very possible that the Northern Empire may have 300,000 oz of mineable resources (not 43-101 compliant). Drilling in 2008 also discovered promising gold veins at the neighbouring Nortoba-Tyson property, also located very close to the mill.

Chapter 10 : Year 2008

Muskeg Tours 2009

Activities of Confederation College

CHIBOUGAMAU DRILLING

A drill rig is no place for a crowd

CON COLLEGE PROGRAMS

On the HEO course, Tom Currie, trainer, checks in with Ted Pile, who is spending his second day operating the huge excavator.

"CONFEDERATION COLLEGE. Education That Works." On November 25, the banner was suspended between shrubs on the north side of Greta Road, four kilometers from Hwy. 584. An access road just over a kilometre long led to the Geraldton gravel pit.

The Greenstone campus of Confederation College is forging ahead with three training programs that help address the downturn in the forest industry and the upturn in mining exploration.

Two joint programs develop skills appropriate to the mining sector, the forestry sector, the construction industry, and the transportation industry. The Heavy Equipment Operator program (HEO) is self-explanatory; the A Z program (AZ) trains drivers for tractor-trailers.

Both programs, begun in October, are 200-hour courses. HEO uses the Geraldton gravel pit as a base; the AZ training vehicle can be observed trundling about the community and local highways. The joint programs have 28 students enrolled.

At the gravel pit, Tom Currie, trainer, said that the second four-week course had begun the day before. The College has several machines leased from local contractors: from Cloutier Contracting, a John Deere excavator (690E-LC) and a rubber-tired loader/backhoe; from Steve Leduchowski Trucking, a D4 Caterpillar bulldozer; from Donald "Bhout" Gosselin, a John Deere 700 bulldozer. From Geraldton Community Forest comes a small office trailer, and from the Municipality, a generator set (to keep machines plugged in when idle) and a portable privy.

Currie was training a student on the D4, his first day on the course. Ted Pile, on Day 2, was operating the excavator. Asked what he thought of the experience so far, Pile said, "I love it!" Activities were confined to the portion of the pit where aggregate had already been extracted. "In four weeks," said Currie, "they'll be comfortable in the seat, and will have done just about everything out here."

Later in the day, Rollie Laroque, the other trainer, and more students would be joining the crew, and Currie would switch to teaching theory.

HEO has 16 students enrolled. Twelve of the 16 students in the HEO will also be taking the AZ. All 12 students in the AZ will be taking the HEO. According to Ken Stevens, Director of the Greenstone and North Shore campuses, the skills of both courses are complementary. For example, a graduate could transport a piece of heavy equipment to a work site, and then operate the machine.

As for HEO graduates, "Only an employer can sign them off," Stevens explained. Therefore, graduates must obtain employment with established contractors in order to obtain certification.

Sam Pytak and Laurier Rossignol are the trainers for the AZ. The program runs a day shift with 3 students, and a night shift with another 3. Already 6 students have finished the AZ. Stevens remarked, "Five out of the 6 got their licences." One is repeating the road test.

The first four Surface Diamond Driller Helper programs graduated a total of 42 students by July, and by press time, the fifth program had wound down after 320 hours, graduating 10 students. For the program, Con College has acquired a drill rig, a drill shack, and a pump shack, and is renting an office/classroom trailer, all located at the west end of Greer Avenue on property owned by the Municipality.

On November 25, the trainer, Ben Desrosiers, was supervising 5 students who, the day before, had been doing placements with North Star Drilling, which has drill rigs operating up the Kinghorn Road, in the Jellicoe area. Another 5 were in the bush. Chibougamau Drilling also was participating in the program.

In the first six weeks, students learned a lot of theory, supplemented with practical training. For the last two weeks of the course, students got to work on operating rigs. Asked what he thought of the experience, student Tom Lennox said, "It was awesome! . . . Everything we learned now makes more sense."

The enrolment in other programs at the Greenstone campus is as follows: 7 students in Practical Nursing; 3 in Pre-Health Sciences; 2 in Office Administration; and 1 in Entrepreneurship.

The Greenstone regional campus has been operating from its Geraldton base for over 30 years. "We are lucky to have such great trainers here in Greenstone," said Stevens. "They bring a tremendous wealth of experience and knowledge to the programs . . . ." Consequently, students have great opportunities locally.

-Times Star

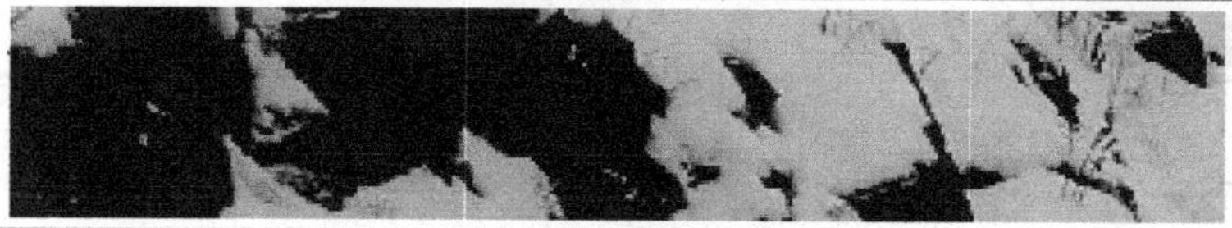

Five of the students on the Surface Diamond Driller Helper course. Ben Desrosiers, trainer, is second from the left. The photographer interrupted the training as they were about the straighten the stovepipe coming from the interior heater of the training rig.

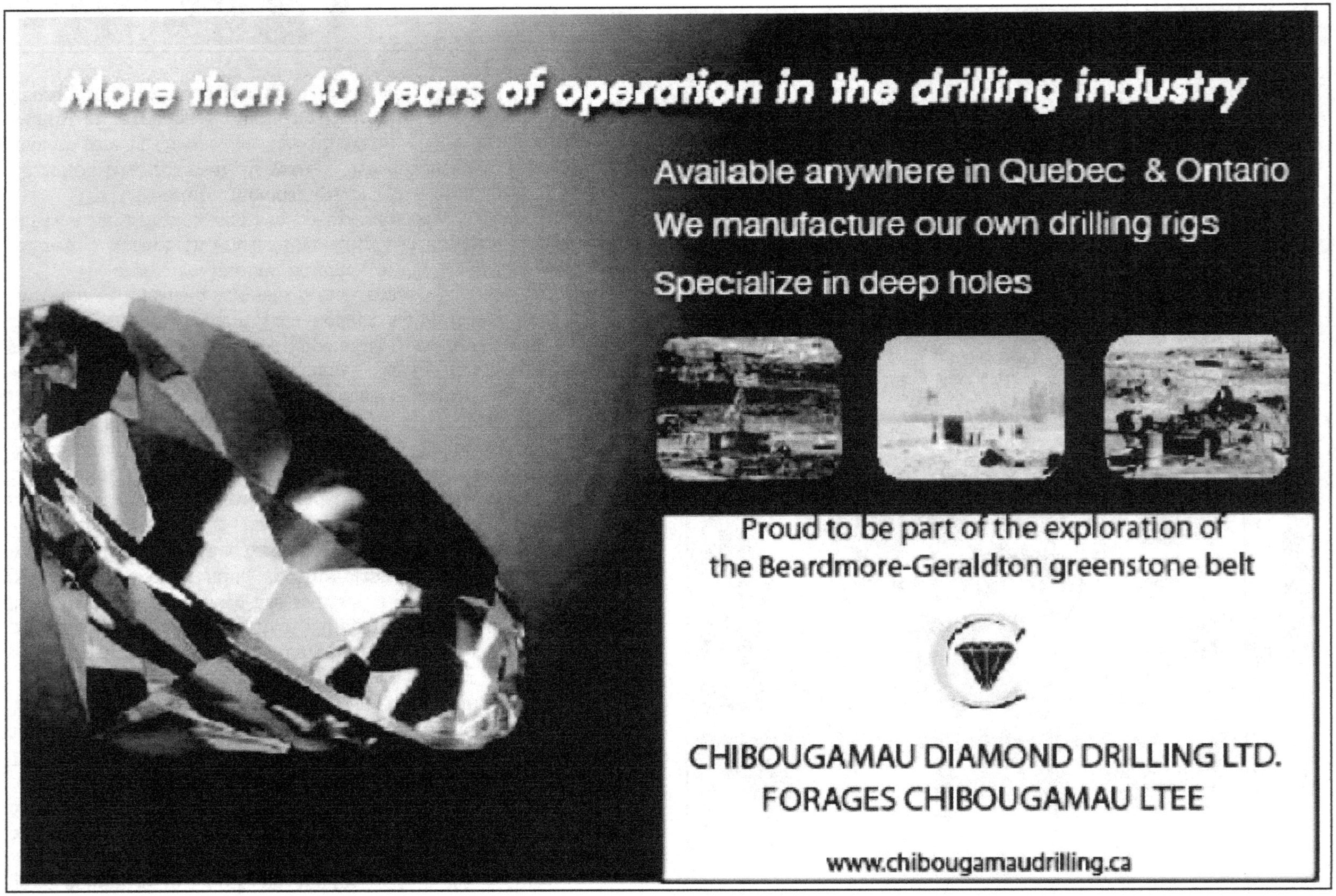
More than 40 years of operation in the drilling industry
Available anywhere in Quebec & Ontario
We manufacture our own drilling rigs
Specialize in deep holes
Proud to be part of the exploration of
the Beardmore-Geraldton greenstone belt
CHIBOUGAMAU DIAMOND DRILLING LTD.
FORAGES CHIBOUGAMAU LTEE
www.chibougamaudrilling.ca

Chapter 11 : Year 2008

Activities of Geraldton & District Chamber of Commerce

Muskeg Tours 2009

High Hill Harbour on Lake Nipigon

CHAMBER ELECTS NEW PRESIDENT

Annual General Meeting a Success

The Geraldton District Chamber of Commerce held its AGM in the basement of St. James Anglican Church on the evening of November 13. It resonated with last year's AGM with its attention to mining themes.

Two guest speakers preceded the dinner, scheduled for 7:00 o'clock. Mayor Michael Power praised the Chamber for its essential role in the community, and for its encouragement of the mining sector. Bruce Hyer, newly elected MP for Thunder Bay-Superior North, delivered the message that he intended to represent everyone in the riding, regardless of political affiliation. He described the small businesses that he himself runs, and said, "I believe the future of our region and of Canada lies in small businesses." He would be asking the NDP Leader, Jack Layton, to assign him the role of critic for small businesses. He concluded by saying he would be paying attention to communities outside of Thunder Bay. "I'm going to be in Greenstone regularly," he said.

A fine buffet-style dinner was served by volunteer members of the church, following which the outgoing President, Peter Kathmann, called the business meeting to order. Kurt Pristanski acted as chair.

The President gave his report. He alluded to the ongoing mining exploration which bode well for the future, and echoed the Mayor's expectation that the forest industry, now in a downturn, will rebound in a different form. He listed the Chamber's program for the past year, referencing the annual trade show, the annual clean-up day, and the buy-local campaign. He noted the successful NOACC semi-annual meeting which the Chamber hosted last spring, which echoed the mining theme established in the Chamber's AGM last fall, and led indirectly to the first annual mining conference last month in Thunder Bay (Prospects North). He said that the shopping bag committee had yet to report, and that the Chamber planned a decorating contest for the holiday season. He concluded by listing the advantages that came with a Chamber membership.

Pristanski chaired the election of officers, with the following result: President, Paula Menchions. Vice President, Karen MacDougall; Executive Secretary, Stephane Parent; Treasurer, Dave Barker.

Peter Kathmann, as outgoing President, would remain on the Executive as a Director. The elected Directors were: Johanna Cloutier, Karen Donio, Margaret Anne Darling, Hugette Venne, Ken Stevens, Julie McMahon, Darlene Lupen, and Diane Starka.

Other positions confirmed were the NOACC Liaison, Julie McMahon (standing in for Katherine Alton); the Municipality of Greenstone Representative, Robert Gendreau; and the Correspondence/Recording Secretary, Kathy Beaupre.

Paula Menchions, newly elected President of the Geraldton District Chamber of Commerce.

The newly elected President, Paula Menchions, made some remarks. She expressed gratitude to the community for her recent experiences: "This community is very welcoming to anyone new." She said that someone had suggested to her a worthwhile tradition to establish in the Chamber, an annual GDCC Business Person of Excellence award.

She read a tribute to the person who has been President of the Chamber for 7 of the last 10 years. "Mr. Peter Kathmann has been a very persistent champion for mining in our area," she said. She cited his exploration work on local properties, and the business he has established for making core racks. "We now have a manufacturing facility here in Geraldton, with four dealerships, distributing product as far away as Yellowknife." She mentioned that Kathmann also ships regularly to Europe such things as ornamental rocks and stones and specialty lumber.

Menchions presented Kathmann with a personal plaque, the master of which will be displayed by the Chamber. Kathmann, somewhat at a loss for words, expressed his gratitude to everyone present.

The conclusion of this meeting will detailed in the Times Star next week. Tim Twomey of Premier Gold, Dave Hunt of Kodiak Exploration, and Maggie Chicoine of The Idea Sculpture each made extensive presentations.

PRESENTERS GIVE POSITIVE SPINS

On the evening of November 13, three presenters suggested reasons to look forward to better times.

Tim Twomey, Exploration Manager, outlined the plans of his company, Premier Gold Mines Ltd. Premier is well capitalized, he said, and "I would say the biggest risk to junior exploration companies today is access to capital." Twomey stated that "the lion's share" of Premier's capital would be spent in the Geraldton area in joint venture with Roxmark Mines Ltd. It is interesting that Premier recently announced a private placement of shares that raised $14 million.

Premier will begin a diamond drilling program in the new year on its newly acquired Barrick Gold property, the site of three former operating mines. Drilling, he said, is a necessary stage to determine an economic resource, i.e., an ore body. In two years, Premier expects that it will go underground, at which time, Twomey said, it would need a lot of capital, in the order of $100 million.

Twomey described the success of Goldcorp in Red Lake Mine, where he played a major role in its new High Grade Zone. "The best place to find a new ore body," he said, "is beside an old one." The newly discovered Kailey Zone, beside the old Little Long Lac mine shaft, looks very promising, and in the next phase, Premier will probe it underground.

Twomey also demonstrated how companies now develop 3-D models using digitized drill hole information. He suggested that such modeling is a useful tool to attract junior companies. "I think you'll attract more exploration here as time goes on," he concluded.

David Hunt of Kodiak Exploration Ltd. described the money his company is sinking into the Greenstone region. Hunt, Regional Exploration Manager, said that Kodiak will have spent $20 million this year alone, mostly on the Hercules Project. The 3-kilometre strike length of the Golden Mile has been tested to 500 metres in depth, for a total of 50,000 m of drill length.

Other Kodiak properties in Greenstone accumulated 19,000 m of drill core this year. Kodiak also conducted an extensive airborne geophysical survey, and employed 15 local people, including First Nations.

As for 2009, Hunt said, "Kodiak is still well financed, with close to $30 million in the bank." The company expects to spend $10 million locally next year. Kodiak will be suspending operations on December 15, and resuming its drilling on February 1.

In conclusion, Hunt pointed out that Kodiak is a junior company, and that its role is to develop mineral deposits to the point where major companies are interested in developing them into mines.

The surprise of the evening was Maggie Chicoine, a motivational speaker. Chicoine's business, The Idea Sculptor, is based in Thunder Bay. The title of her presentation was "Waiting in the Wings". Anyone who tries to pin down the vision and high-flying ideas of a motivational speaker is just asking to be criticized. But here goes:

Chicoine began by throwing out three statements, and asking for interpretations. One, "The last thing a fish discovers is water." It is true that we do not see ourselves, and our environment, as others do. Two, "Complacency is the safe spot." We are usually inclined to accept our lot, and to wait for relief. Three, "You got to be very careful if you don't know where you're going, because you might not get there." (Yogi Berra) As Chicoine summed up the interpretations offered up, "You need to have a vision and a clear plan so people can be part of it."

With respect to our situation here in Greenstone, we are oblivious to the attractions we have; we are undergoing an economic downturn; and we are preparing to wait it out, or hoping the government will bail us out. And we are not taking advantage of the one bright star in our firmament: mining exploration. "Until people say this is part of who I am, they're not going to get it."

Some of the evening's presenters, L to R: Tim Twomey (Premier Gold), Dave Hunt (Kodiak Exploration), Maggie Chicoine (The Idea Sculptor), and Bruce Hyer (MP, Thunder Bay-Superior North).

The presenter explored three more concepts. Using a necktie to demonstrate, she showed we can depict our history as a series of knots. Each knot was a period of economic activity, associated with, for example, the mining era, and the forestry era, and currently, the economic slump. Until we extricate from each successive knot – until we move on – we cannot take advantage of new opportunities.

Second, she asked how we see ourselves. Do we see ourselves as a vulnerable community, or as a resilient one? Are we victims, blaming others, cranky? Or are we ready to create our future by our own initiatives?

Third, she suggested we look around us, and see, perhaps for the first time, something unique we can offer or sell. She used a number of examples of communities who did just that. Manitouwadge now sells extreme adventures. Grand Marais has transformed into a premier tourist mecca. Nipigon is known far and wide for its iconic "Nipigon nylons" (just Stanfield woolen socks), and is incorporating the icon into community projects. Kenora has "Kenora dinner jackets". Nakina has a parallel, the "Nakina dinner jacket". But Kenora now has a business which sells its "dinner jackets" (off-the-rack woolen red-and-black-checkered coats) for $99 apiece. "They're not waiting in the wings," said Chicoine, "waiting for someone to come and help them."

Chicoine concluded by suggesting that the current economic downturn is an opportunity for the community to plan a strategy for a brighter future.

As the annual general meeting of the Geraldton District Chamber of Commerce adjourned, the close to 50 people in attendance left with smiles on their lips and springs in their steps.

-Times Star

Lunch break on the Gripp River

Chapter 12 : Year 2008

Muskeg Tours 2009

Activities of Ontario Geological Survey

Superior Canadian Resources Inc. Photo

A BANNER YEAR IN THE GREENSTONE AREA

The 2008 season saw a record expenditure of over $35 million by thirteen exploration companies in the main Beardmore – Geraldton Belt. This encompasses a region stretching from Lake Nipigon in the west, through Geraldton and Longlac, to Klotz Lake in the east. The entire greenstone belt was covered by claims during a staking rush initiated in October of 2007, following an announcement by Kodiak Exploration Limited of a major gold discovery on their Hercules Property in Elmhirst Township. Extensive airborne surveys were conducted, a significant number of new and historical gold showings and occurrences were stripped, trenched and sampled, and over 160,000 meters of diamond drilling was completed.

Kodiak Exploration Limited, through numerous option deals with local prospectors and extensive claim staking, became the largest land holder in the area and control approximately 1400 square kilometres across the belt. The company's success attracted national attention and a core of other junior exploration companies with strategic land positions in the area. These included Premier Gold Mines Limited, Sage Gold Inc., Alto Ventures Ltd., Roxmark Mines Limited, Mantis Mineral Corp., Laurion Mineral Exploration Inc. and Melkior Resources Inc. Premier, in a joint-venture agreement with Roxmark and a recent acquisition from Barrick Gold Corporation, is actively exploring five past-producing gold mine properties in the Geraldton area. Ontex Resources Limited, already established in the west-central portion of the main belt in Irwin Township, holds a 35 km long claim package hosting a known National Instrument 43-101-compliant gold resource of 694,700 ounces (indicated and inferred) at the Brookbank deposit. The company has continued to build on this resource during the year with an ongoing drilling program.

2008 will be remembered as a banner year in the Greenstone area, with unprecedented levels of exploration activity and expenditures. Not since the original exploration rush to the area in the 1930's has the area's potential attracted so much interest and investment.

- Gerry White, *P. Geo.*
District Geologist
Ontario Geological Survey

FACSIMILE EDITION ENDS

ABOUT THE AUTHOR

Edgar J. Lavoie lives in Greenstone, Ontario, where he retired after 35 years' teaching. He and his wife, Olga, have two children, who in turn have families of their own. The author has solid credentials in writing and publishing, ranging from little magazines, news articles, and short stories to local history books. He published his first novel, ***The Beardmore Relics,*** in 2011, and his second, ***Geraldton Back Doors,*** in 2013. Both novels feature the geography, history, and character types of his beloved region. He and Olga live in the boreal forest north of Thunder Bay, in a big cabin on a big lake. ***Bush Histories I*** is the first book of a new series that features Northwestern Ontario.

AUTHOR'S WEBSITE
www.WhiskyJackPublishing.ca

AUTHOR'S BLOG
http://ejlavoie.wordpress.com

AUTHOR'S FACEBOOK PAGES
Edgar J. Lavoie, Author
Edgar Lavoie
Greenstone History

Made in the USA
Monee, IL
08 November 2026

67384901R00111